ENVIRONMENTAL AND WATER RESOURCES HISTORY

PROCEEDINGS AND INVITED PAPERS
FOR THE ASCE 150TH ANNIVERSARY (1852–2002)

November 3–7, 2002
Washington, DC

SPONSORED BY
Environmental and Water Resources Institute of the American Society of
Civil Engineers

EWRI National History and Heritage Committee

EDITED BY
Jerry R. Rogers
Augustine J. Fredrich

ASCE *American Society of Civil Engineers*
1801 ALEXANDER BELL DRIVE
RESTON, VIRGINIA 20191–4400

Abstract: *Environmental and Water Resources History* consists of papers presented at the Civil Engineering Conference and Exposition held in Washington, DC, November 3-7, 2002. These proceedings bring together for the first time a series of papers that focus exclusively on various aspects of environmental and water resources history. The proceedings papers address such topics as environmental engineering history and developments, hydraulic engineering pioneers, bureau of reclamation history and developments, university water and hydraulic education/research, hydrology and water resources planning, invited river history summary, and EWRI formation and history of ASCE technical divisions and codes/standards activities.

Library of Congress Cataloging-in-Publication Data

Environmental and water resources history : proceedings and invited papers for the ASCE 150th anniversary (1852-2002) : November 3-7, 2002, Washington, DC / sponsored by Environmental and Water Resources Institute of the American Society of Civil Engineers, EWRI National History and Heritage Committee ; edited by Jerry R. Rogers, Augustine J. Fredrich.
    p. cm.
    Includes bibliographical references (p.).
    ISBN 0-7844-0650-2
    1. Water resources development--United States--History--Congresses. 2. Hydraulic engineering--United States--History--Congresses. 3. Environmental engineering--United States--History--Congresses. 4. Water-supply--United States--Planning--History--Congresses. I. Rogers, Jerry R. II. Fredrich, Augustine J. III. Environmental and Water Resources Institute (U.S.)

TC423 .E58 2002
333.91'00973--dc21   2002034154

Foreword

These first *Environmental and Water Resources History Proceedings* of the Environmental and Water Resources Institute (EWRI) of the American Society of Civil Engineers (ASCE) are dedicated to our wives and families: Cecelia and Laura, Joseph and Gregory and their families, and Donna and Sharon and Steven and their families plus Don W. Rogers, Jr. We thank all of them for supporting us in our ASCE activities for more than 30 years.

These Proceedings are the co-editors' and authors' contributions to the ASCE 150th Anniversary observance. They are intended to bring together for the first time a series of papers that focus exclusively on various aspects of the history of environmental and water resources subjects. (Some environmental and water resources papers were published in the *International History and Heritage Proceedings*, ASCE, October 2001.) Through our service on ASCE's National History and Heritage Committee (HHC), we became aware of the absence of a coordinated collection of papers on water/environmental history and heritage. The creation of a water/environmental program track with a history focus for the November 3–6, 2002 ASCE Conference and 150th Anniversary in Washington, D.C. provides a nucleus for such a collection. These Proceedings include papers from the technical sessions comprising that track and invited papers outlining the history of various governmental and academic organizations that have played important roles in the nation's water resources and environmental activities. In particular, the Bureau of Reclamation of the U.S. Department of the Interior celebrated its Centennial (1902–2002) with a Reclamation History Symposium June 17–19, 2002. Several papers from that Symposium were invited and contributed to this ASCE 150th Anniversary Proceedings. We thank the Bureau of Reclamation for its participation in the 2002 ASCE National Conference and these Proceedings and for the cover photograph of the upstream view of Hoover Dam during the filing of Lake Mead around 1936.

Glenn Brown of Oklahoma State University is organizing an EWRI "Darcy Memorial Symposium" June 24, 2003 for the 2003 EWRI Congress in Philadelphia. The new EWRI History and Heritage Committee has proposed a second Environmental and Water Resources History Symposium with Proceedings for the summer 2004 EWRI Congress in Salt Lake City. (Members of that committee have been helpful in contributing papers or contacts for invited papers for these Proceedings, and we are grateful for their assistance.) These forthcoming activities offer the prospect of additional publications to further document the history and heritage of this important aspect of the civil engineering profession.

Finally, we are grateful to the University of Arkansas Civil Engineering Department for our undergraduate foundation in civil engineering and water resources/environmental engineering. In particular, we thank Professor Loren Heiple (past Department Chair) for his encouragement and for his efforts in the formation of the careers of many civil engineering graduates of the University of Arkansas.

-- Jerry R. Rogers and Augustine J. Fredrich, Co-editors

Contents

Hydrology and Water Resources Planning

Invited River History Summary

*EWRI Formation and History of ASCE Technical Divisions
and Codes/Standards Activities*

A History of Environmental Engineering in the United States

William C. Anderson, P.E., DEE[1]

Environmental engineering is a relatively new name for a type of engineering that began in the United States in the 1830s. Under different names, it continued to evolve to satisfy environmental challenges posed by urbanization, suburbanization, and the other needs of the nation during the industrial revolution in the late 1800s through the information revolution of the 1990s.

Until 1970, when Earth Day captured the public's attention leading to a concentrated effort to clean up the environment, the profession was practiced by but a few. Yet, the pioneers — Mills, Chesbrough, Sedgwick, Hazen, Metcalf, Eddy, Camp, Fair, Wolman, to name a few — blazed a trail establishing design protocols still in use today. Of necessity, this section is but a brief overview of the rich heritage on which modern environmental engineering is founded.

The Beginning

Hydraulic engineering best describes environmental engineering at its birth. Early communities were usually located on or adjacent to plentiful sources of fresh water. As these communities grew and people were forced to live farther and farther from the water source, private companies formed to convey water to the outlying areas. By 1800, there were 18 private waterworks in the U.S. (McKinney 1994).

The inability of these private water companies to meet the water needs of rapidly-growing cities forced the larger municipalities such as New York, Boston, and Chicago to consider public water systems. Colonel De Witt Clinton, Jr., an Army engineer and son of the former governor of New York, was retained to examine New York's water supply needs in 1832. He recommended that water be obtained from the Croton River and conveyed to New York City through a 40-mile aqueduct. In 1836, John B. Jervis, a self-educated hydraulic engineer who had learned his engineering on the Erie Canal and the Mohawk and Hudson Railroad, began construction of a dam across the Croton River to provide storage for the aqueduct

[1]Executive Director, American Academy of Environmental Engineers, 130 Holiday Court, Suite 100, Annapolis, MD 21401; phone 410-266-3311

(Jervis 1876). By 1842, fresh water began flowing from the Croton Reservoir to New York City through the aqueduct which carried 95 Mgal/day (4.16 m3/sec) (Hazen 1907).

In Boston, city officials, noting the success of the Croton Aqueduct in New York, retained Jervis and Professor Walter R. Johnson of Philadelphia to find a new water supply for Boston in 1844 (Brandlee 1868). The team identified Lake Cochituate as the best source of water and Jervis designed the Cochituate Aqueduct. At the same time, Ellis S. Chesbrough was retained as Chief Engineer of the West Division of the Boston Water Works (Cain 1991). Under Chesbrough, construction of the Cochituate Aqueduct was completed in 1848 and Boston established its public water system. He became the Commissioner of the Boston Water Works and Boston's first City Engineer.

As City Engineer, Chesbrough was responsible for managing the stormwater sewer system. Boston had taken over the storm sewers in 1823 to ensure that they were properly maintained and that future sewers were built to proper specifications (MSBH 1903). Storm sewers were designed to flow by gravity to the nearest stream and eventually to the ocean. Because Boston prohibited the dumping of human sewage into storm sewers, sanitary sewage was collected in privies and vaults which were pumped out at regular intervals. This septage was carried in special wagons to farms outside the City. But, with the increased water supplies, more water closets and indoor bathtubs increased the flow of wastewaters and quickly filled the sewage storage vaults. A few homeowners solved their problem by building their own sewers and connecting them to the storm sewers without official permission. In addition to domestic sewage pollution of storm water, horse manure from the streets ended up in storm sewers when it rained. While smaller communities typically constructed two sets of sewers, a storm sewer and a sanitary sewer, Boston, like New York and other large cities, combined the sanitary sewers and the storm sewers in one pipe for economic reasons. Under either approach, sewer design required hydraulic engineering.

After serving Boston as its City Engineer for seven years, Chesbrough resigned in 1855 and moved to Chicago. There, he developed its sewerage system, the accomplishment for which he is most remembered. Chesbrough developed an innovative (for the time) plan to employ combined sewers to drain the city's waste into the Chicago River which required raising the elevation of most of the downtown area. The new sewage system contaminated Lake Michigan, the city's water supply. To correct this problem, Chesbrough developed and implemented an intake works offshore which was connected to the city by a 2 mile-long tunnel.

A Discipline Takes Shape

As the fruits of hydraulic engineering water systems serving the public became more common, and concerns about the quality of the water used for public water supplies grew, Sanitary Chemistry emerged as a new facet of environmental engineering. In 1873, the Massachusetts State Board of Health asked Professor William Ripley Nichols, who was in charge of Chemistry at the then infant Massachusetts Institute of Technology (M.I.T.), to analyze the water quality of the major rivers used for public water supplies in Massachusetts (MSBH 1874). Nichols agreed and set up the first Sanitary Chemistry Laboratory in 1874 to perform water analyses. Ellen Richards, then Ellen Swallow, the first coed at M.I.T., had just graduated with her BS degree in Chemistry from M.I.T. She was named Professor Nichols' assistant and did most of the analyses at the new Sanitary Chemistry Laboratory. She went on to become one of the foremost sanitary chemists in the United States.

Water pollution problems also increased as populations rose and industrial and water system development advanced. Research by the Sanitary Chemistry Laboratory at M.I.T. into the chemical quality of Massachusetts' river water showed that pollution was becoming significant and needed to be reduced before wastewaters contaminated public water supplies. Because there was no viable system for treating sewage, in 1887 the Massachusetts State Board of Health established the Lawrence Experiment Station, the first of its kind, to do the necessary research. Hiram F. Mills, a hydraulic engineer from Lawrence, Massachusetts, who also was a member of the Board of Health, was chosen as the station's first director. Mills recognized that research on wastewater treatment required not only engineers, but also chemists and biologists. As a member of the M.I.T. Board of Trustees, Mills did not hesitate to draw upon M.I.T. to supplement the staff at the experiment station. Professor William T. Sedgwick was appointed as Consulting Biologist and Dr. Thomas M. Brown, Professor of Chemistry, was appointed Consulting Chemist to the State Board of Health (McCracken and Sebian 1988).

In the beginning, environmental research at the Lawrence Experiment Station relied heavily on British research on intermittent sand filtration. In Britain, Sir Edward Frankland had demonstrated that intermittent sand filtration was a viable method for treating municipal sewage. Picking up on his work, Allen Hazen, George Fuller, and Harry Clark, the staff of the Lawrence Experiment Station, demonstrated that wastewater treatment was a biochemical process in addition to a physical process. The station's results, published in two large volumes in 1890 (MSBH 1890), were carefully studied across the world and stimulated considerable research on municipal wastewater treatment. These reports, Purification of Sewage and Intermittent Filtration of Water and Examination of Water Supplies and Inland Waters of Massachusetts, contained over 1,500 pages because of Mills' insistence that all the data collected be shared with others (McCracken and Sebian 1988). Using the Lawrence Experiment Station results, the British developed trickling filters. By 1900, biological wastewater treatment concepts were well established.

The findings by the Lawrence Experiment Station were made possible by the newly-discovered discipline of bacteriology. The discipline rapidly advanced after Dr. Robert Koch, a German physician and self-educated bacteriologist, demonstrated that bacteria could cause diseases (Collard 1976). Koch became a university professor in Berlin and under his leadership, research in the fledgling field moved quickly from growing bacteria on potato slices to gelatin and agar during the 1880s. This development enabled Professor Sedgwick and Edwon O. Jordan, a recent graduate student of Sedgwick's, to conduct research on sewage bacteriology.

Their efforts focused on isolating and identifying different bacteria in pure cultures. They discovered that bacteria isolated from intermittent sand filters used to treat municipal sewage were non-pathogenic. However, isolating pathogenic bacteria was difficult, even from contaminated wastewaters. Professor Sedgwick, Jordan, and Ellen Richards demonstrated that nitrification was a biological process that could be caused by bacteria, but they were unable to isolate the nitrifying bacteria in pure culture. Nitrate formation was the primary indicator for the stability of wastewaters after filtration. Allen Hazen, the first Chemist-in-Charge at the Lawrence Experiment Station, suggested using a liquid medium containing only inorganic materials rather than using gelatin. After several attempts, Jordan developed a nitrifying culture in an organic-free medium. At the same time, in Russia, Serge Winogradsky, a soil microbiologist, successfully applied the same approach. Winogradsky was able to isolate nitrifying bacteria on silica gel media.

Professor Sedgwick's studies also included algae and protozoa found in surface waters. Sedgwick developed a quantitative method for counting thee large microorganisms that was modified by George W. Rafter. The Sedgwick-Rafter Concentrator and Counting Cell became standard equipment for evaluating algae and protozoa in water supplies. The work at M.I.T. and the Lawrence Experiment Station established Water Microbiology as a major area for environmental engineering.

The concept that bacteria cause disease was not readily accepted in 1890, despite its wide acceptance in the scientific community. Even the isolation of several disease-producing bacteria in pure cultures was not considered adequate proof. A typhoid fever epidemic in Lowell and Lawrence, Massachusetts, in December 1890 and early 1891 provided the impetus for a concerted research effort.

By using engineering techniques and scientific logic, Sedgwick developed a form to collect data on a house-to-house basis to determine the source of infections. Together with George V. McLauthlin, Sedgwick began collecting data from every house reporting cases of typhoid fever and those in the immediate vicinity. Each house with a reported death was marked on a map of Lowell along with the houses with reported typhoid cases. This mapping showed that some areas of the city were affected more severely than others. This data, when compared with a map of Lowell's five different water systems, clearly demonstrated that the areas that obtained drinking water from the Merrimack River were most affected by the typhoid fever epidemic. Applying their methodology in Lawrence, Sedgwick and McLauthlin found that people who used Merrimack River water were most affected. The documentation of these epidemics of typhoid fever, together with several other smaller epidemics transmitted by water, milk, and direct contact, provided irrefutable proof that typhoid fever could be spread by polluted water.

Formal engineering design evolved slowly as engineers learned the best design concepts, but it eventually became the backbone for environmental engineering. Early design efforts were focused on water distribution and sewers. After the Civil War, there was a decades-long debate between proponents of "separate" versus "combined" sewers. Some engineers favored building separate sewers for stormwater and sanitary wastewater. Others favored combining stormwater and wastewater into a single sewer. George E. Waring was among the proponents of separate sewers. Waring, a self-educated engineer, became interested in sewerage systems in the 1870s. He was retained by the city of Memphis, Tennessee, as a consulting engineer after epidemics broke out there in 1878 and 1879. Waring recommended that the city construct separate sewers, using small-diameter pipes with automatic flush tanks (Odell 1880). After the Memphis project, Waring worked on the Buffalo, New York, trunk sewer and was also appointed to the National Board of Health. In 1895, he was named Commissioner of Sanitation in New York City. In three short years, Waring improved the city's solid waste collection and processing which had been one of the worst of the major cities.

James P. Kirkwood was one of the first American engineers to design a slow sand filter for water treatment based on data collected in Europe as part of a study of water treatment for St. Louis, Mo., published in 1869 (Baker 1948). Yet, Kirkwood's initial design of slow sand filters for St. Louis was ignored. It was not until 1871 that Poughkeepskie, New York, constructed a slow sand filter based on his design.

In response to the typhoid epidemics of 1890 and 1891, Hiram Mills designed an intermittent sand filter to treat Merrimack River water in Lawrence. The sand filter was placed into operation in 1893 and removed 98 percent of the bacteria from the polluted river water (MSBH 1894). Deaths from typhoid fever took a dramatic drop, clearly demonstrating the value of sand filtration for purifying polluted water. With slow sand filters proven, considerable effort was directed toward developing mechanical, rapid sand filters. However, the interest in patenting such devices slowed their development.

Rudolph Hering made perhaps the most significant early contributions to the development of engineering design for water supply and sewage treatment. Hering was born in Philadelphia and educated in Germany. He returned to the United States as an engineer, eventually becoming the Assistant City Engineer of Philadelphia in 1873. In 1880, Hering was commissioned by the National Board of Health to go to Europe and study the latest methods for sewage treatment. He presented a report on his findings to the American Society of Civil Engineers in 1881 (Hering 1881). In 1889, he was appointed by President Harrison to make plans for sewerage and drainage in Washington, D.C. Over the years, Hering prepared water supply and treatment studies for 150 cities. It is not surprising that he became known as the "Dean of Sanitary Engineering" (ASCE 1972).

While working at the Lawrence Experiment Station, Allen Hazen had examined chemical precipitation and the sedimentation processes. Because of this expertise, Hazen was invited to Chicago in 1893 to operate the wastewater treatment plant constructed for the Columbian Exposition, a showcase for modern technology. A sewerage system connected each of the exposition's major buildings to the wastewater treatment plant enclosed in a separate building. Because of the flat terrain, each building was equipped with an ejector to lift the sewage into the sewers and convey it to the treatment plant. Unfortunately, the multitude of ejectors and their random operation created a number of problems such as periodic sewer ruptures caused by system pressures exceeding pipe capacity and peak flows disrupting the settling characteristics in the 30 foot high Dortmund settling tanks. Coping with the Columbian Exposition wastewater treatment plant operations showed Hazen that there was a major difference between operating pilot plants at the Lawrence Experiment Station and actual full-scale plants. When the Columbian Exposition closed, Hazen returned to Boston to become a consulting civil engineer, eventually locating his practice in New York City. His paper, "On Sedimentation", was presented before the American Society of Civil Engineers in 1904 and became one of the classic papers on sedimentation theory and design (Hazen 1904). In 1907, he summarized water treatment plant design in his book, Clean Water and How to Get It (Hazen 1907). Hazen became one of the most successful and respected water treatment plant design engineers in the early 1900s.

As Hazen had demonstrated, design engineering grew out of need and experience rather than from theory. This was true for both sewage treatment plants and water treatment plants, although there was more interest in water treatment plants. Most design engineers depended on information published in the latest engineering magazines. In the United States, engineers quickly adopted trickling filter designs and activated sludge technology from the British and Imhoff tanks from the Germans. As each new plant was constructed, design engineers learned what worked and what did not. Leonard Metcalf and Harrison P. Eddy, consulting engineers in Boston, brought the best of American wastewater technology together for all engineers in 1915 with the publication of American Sewerage Practice (Metcalf and Eddy 1915). They set a standard for professionalism while demonstrating their knowledge for future customers. They also provided texts for teaching future generations of design engineers.

A discipline is born when the development of knowledge and its application evolves from individual experimentation into a formal course of study. In 1889, M.I.T. established the first program in Sanitary Engineering (Wylie 1975). It was designated Course XI and incorporated courses in sanitary chemistry and sanitary biology into the Civil Engineering Department. The new department was named Civil and Sanitary Engineering, and degrees were offered at the undergraduate level in Civil Engineering and in Sanitary Engineering. By 1893, when engineering faculty from across the country gathered at the Columbian Exposition in Chicago to organize a professional society to represent engineering educators, the only other school that has a degree program in Sanitary Engineering was the University of Illinois, which offered a degree in Municipal and Sanitary Engineering. A survey of engineering education in

1899 by Ira O. Baker, President of the Society for the Promotion of Engineering Education (SPEE) (Baker 1900), indicated that there were 110 engineering schools, but only 89 were active and only two schools offered degrees in Sanitary Engineering. There were 9,679 students enrolled in engineering, with 19 in Sanitary Engineering. Of the 1,413 engineering degrees awarded in 1898-1899, only one was in Sanitary Engineering (McKinney 1994).

The 20th Century Before World War II

Environmental engineers entered the 20th century with hope and aspiration of the opportunities that lay ahead. M.I.T.'s Sedgwick was confident that Sanitary Engineers had a special place in the future of technology, even though most Course XI graduates at M.I.T. were still concentrating on hydraulic engineering. Environmental engineering was water-oriented. The need for safe water supplies for a dynamic, growing nation occupied many. Additionally, with connection between safe drinking water and polluted waters firmly established, substantial effort was focused on the abatement of water pollution. Interestingly, even air pollution, a post World War II focus, attracted some interest.

While these various facets of environmental engineering are interconnected, technology development and applications were pursued separately. Any interchanges between the specialties existed primarily at universities teaching sanitary engineering, the Public Health Service, and associated state departments of health.

George C. Whipple, who had been a biologist at the Boston Water Works for eight years after graduating and a member of Hazen, Whipple and Fuller, water and wastewater consultants in New York City since 1903, was appointed Professor of Sanitary Engineering at Harvard University in 1911. After his appointment, he joined forces with his professor Sedgwick at M.I.T. and in 1913 the M.I.T.-Harvard School of Public Health was established with Sedgwick as the Program Head and Whipple as Secretary. This association enabled the Harvard Sanitary Engineering program to maintain a focus on Public Health Engineering in addition to customary civil engineering.

The Harvard Sanitary Engineering faculty was joined in 1918 by one of its recent graduates, Gordon Maskew Fair, an immigrant from South Africa via Germany. This inauspicious beginning launched a career that would have a profound impact on the profession. A survey taken in 1949 showed that about half of all American doctorate degrees in Sanitary Engineering up to that time had been earned at Harvard, and over half of the State Sanitary Engineers had received advanced degrees under Fair's direction (Anderson 1986).

When Fair began work, the technology of most processes and practices for environmental control were characterized by a high degree of empiricism. The remedies prescribed for freeing air and water of pollutants were derived from workaday experience rather that from scientific observation and analysis. His research in environmental science was motivated by the belief that a set of theoretical principles governed a wide range of artificial and natural purification processes — and that these could be specified in mathematical language so that engineers could use an orderly process of calculation in designing treatment works for water and air. His success in developing the theory of purification kinetics was embodied in widely read books and papers including his well-known textbook, Water Supply and Waste Water Disposal (with J.C. Geyer). He provided additional contributions in limnology, the broad application of the principles of physical chemistry to complex processes of water purification, to specific problems of quantitative measurement of tastes and odors, and mechanisms of biodegradation of certain organic compounds in streams. Perhaps his greatest achievements stemmed from his ability as a theorist to deploy the methods of science and the techniques of mathematical analysis, a precursor of today's emphasis of mathematical modeling and the key to computerization, to a discipline that had evolved for many centuries as a practical art. Fair's grasp of environmental engineering was visionary and prophetic. Years ago he understood that environmental control is a multi-media challenge.

The impact of sewage on water quality and the need for safe drinking water identified in the Massachusetts and other major urban centers in the U.S. gave impetus for the creation of a sanitary engineering component within the US Public Health Service (PHS) in 1890. At the start of the new century, this predecessor of EPA came to be a dominating force. In 1901, Congress authorized the construction of a PHS Hygienic Laboratory, "for the investigation of infectious and contagious diseases."

This was followed by commissions in 1908-1909 to study water pollution and protect water supplies in Lake Michigan and Lake Erie. A major reorganization in 1912 gave the PHS a broad mandate to study the diseases of man and conditions influencing their propagation.

In 1913, a group of medical officers, engineers, and scientists took over the laboratories at an abandoned Marine Hospital in Cincinnati, Ohio, with a mission to control water pollution. This center produced much of the fundamental research on which the control of water pollution is founded, including:

- definition of the Oxygen Sag Equation by Streeter and Phelps;
- confirmation of the rate of oxygenation of polluted waters by Theriault;
- confirmation of the rate of atmospheric reaeration by Streeter;
- definition of the elements of bacterial pollution by Hoskins;
- development of major elements of stream biology by Purdy; and
- initiation of studies on industrial wastes (Dworsky 1990).

For nearly thirty years, between 1913-1938, many of the Cincinnati group, augmented by a second, but still small, wave of engineers and scientists, structured and implemented plans which carried the nation rationally toward its goal of water pollution control. Some of these include:

- a strategic selection of rivers to understand the properties of their differences (1914);
- the initiation of a comprehensive survey of stream pollution (1915);

- support for the growth and improved capacity of the states to participate in efforts to control stream pollution (1920);
- public education efforts focusing on the importance of water pollution control measures as an aspect of comprehensive water resources development (1936); and
- increased technical assistance to states through the creation of a separate PHS Office of Stream Sanitation (1932) (Dworsky 1990).

Their efforts culminated in Public Law 845, enacted by the 80th Congress on June 30, 1948, the nation's first comprehensive Water Pollution Control Act.

While educators and government officials labored to understand the science of water pollution control in the first half of the century, wastewater was being treated in urban areas of all sizes. Technology developments continued to flow back and forth across the Atlantic. The studies at Lawrence Experiment Station fueled the creation of the trickling filter in England and the Imhoff Tank in Germany. These developments were applied in many U.S. communities; Chicago used Imhoff tanks to treat over 500 million gallons per day by the mid 1920's. In 1914, the activated sludge method of wastewater treatment, today's most commonly-used wastewater treatment technology, was developed in Manchester, England. Shortly thereafter, in 1918, Houston, Texas, placed the first large-scale activated sludge plant in the U.S. in operation.

The work at the Lawrence Experiment Station fostered not only wastewater treatment, but water treatment as well. The station's research proved the efficacy and value of sand filtration to protect human health. First with slow sand filters and then into the 20th century with rapid sand filters, water filtration became the preferred treatment technology for all but those with a pure, upland, source of supply.

The Interstate Quarantine Act in 1893 laid the foundation for the control of communicable disease and the regulation of its carriers, such as water. A regulation pursuant to this act compelled trains and other interstate carriers to use waters of known quality which had been certified by the local health authority. The nation's first drinking water standards were adopted in 1914 as an aid to the enforcement of the Quarantine Act. These were implemented by state agencies with support from the PHS.

Arguably, the most significant development in water treatment in the first half of the 20th century occurred in 1923. That was the year Abel Wolman, who had recently been appointed Chief Engineer of the Maryland Department of Health, developed and perfected techniques for controlled chlorination of water. These techniques made possible the prescription of chlorine feed rates for water leaving the treatment plant sufficient to provide for effective and reliable disinfection of the water supplied. With filtration and/or reliable disinfection, the major public health threat of water-borne disease ceased to exist in the U.S. by World War II.

Clearly, water-related issues occupied the first half of the 1900s. However, air quality concerns began to receive attention shortly after the turn of the century. The major atmospheric concern stemmed from the presence of smoke. Up to the late 1940s, most American urban centers had smoke abatement agencies. The transfer of interest from solely smoke abatement to more comprehensive air quality issues came about as a result of

developments in the field of industrial hygiene. Early atmospheric pollution studies were an extension of the science of industrial hygiene and included:
- a PHS and US Bureau of Mine study of silicosis (1910)
- organization of a Division of Industrial Hygiene by the PHS (1912).
- studies of air in industrial workshops, e.g., carbon monoxide from the use of gas-heated equipment (1916); and
- studies by the PHS of municipal dust, the radium dial painting industry, and a comparative study of air pollution in fourteen of the largest cities (1913) (Dworsky 1990).

Not withstanding the foregoing, the seminal event in the speciality of air pollution control was triggered by an air pollution episode in Donora, Pennsylvania, in 1944; twenty persons died and 5,910 became ill. It triggered the first major comprehensive study of air pollution by the PHS. The resulting Surgeon General's Report noted:
- "This study is the opening move in what may develop into a major field of operation in improving the nation's health.
- We have realized, during our growing impatience with the annoyance of smoke, that pollution from gases, fumes and microscopic particles was also a factor to be reckoned with.
- The Donora report had completely confirmed beliefs we held at the outset of the investigation:
  1.   how little fundamental knowledge exists regarding the effects of atmospheric pollution on health, and
  2.   how long range and complex is the job of overcoming air pollution."

Post World War II

The industrial explosion that accompanied World War II had two significant impacts in environmental engineering. On the one hand, the richness and diversity of technological exploitation resulted in a plethora of chemical and other industrial wastes discharged to the water and air and deposited on the land. At the time, these were of little concern to a public tired from war and anxious to enjoy the fruits of peace. They systematically ignored the concerns and warnings from environmental engineers of the day. But the day-by-day, year-in-year-out callous disregard of wasteful production sowed the seeds for an environmental revolution triggered by Rachel Carlson's 1964 book, Silent Spring, and which culminated with Earth Day 1970. These events unleashed a torrent of regulations which has dominated the profession ever since.

The immediate post World War II period was also marked by a significant increase in environmental engineering research, made possible by significant federal grants to universities. A pattern was established and continues to this day that has significantly increased the knowledge of the science underlying environmental engineering. These grants also produced increasing numbers of environmental engineers with masters and doctorate degrees and spawned a new class of environmental engineer – the academic-researcher – which replaced, for the most part, the academic-practitioner who was the norm before the war.

As the next millennium approaches, environmental engineers practice in a world of increasing complexity as they attempt to work in harmony with a most complex system - nature - and a rapidly expanding knowledge base made possible by the "information age". Yet, it is meaningful to reflect upon the profession's history and to take note that what environmental engineers do today and tomorrow is grounded on fundamental principles developed and perfected 50, 100, and 150 years ago. These principles remain those which today's environmental engineers must know to be deemed minimally competent and, therefore, entitled to a license to practice the profession.

References

American Society of Civil Engineers, Committee on History and Heritage. 1972. *A Biographical Dictionary of American Civil Engineers*, ASCE Historical Publication no. 2.

Anderson, W.C. 1986. "Dr. Gordon Maskew Fair - The Master", *The Diplomate* vol. 22 no. 3, pp 6-8.

Baker, I.O. 1900. "Engineering Education in the United States at the End of the Century," *Proc. 8th Annual Meeting of Soc. for Prom. of Engr. Educ.*, vol. 8, pp 11-27.

Baker, M.N. 1948. *The Quest for Pure Water,* Amer. Water Works Assoc.

Bradlee, N.J. 1868. *History of the Introduction of Pure Water into the City of Boston*, Alfred Mudge & Sons, Boston, Mass.

Cain, L.P. 1991. "Raising and Watering a City: Ellis Sylvester Chesbrough and Chicago's First Sanitation System," The Engineers in America, pp. 69-88, Univ. of Chicago Press, Chicago, Ill.

Collard, P. 1976. The Development of Microbiology, Cambridge Univ. Press, Cambridge, Mass.

Dworsky L.B. 1990. "The United States Public Health Science", *The Diplomate*, vol. 26, no. 4, pp 7-12, 31.

Hazen, Allen. 1904. "On Sedimentation," *Trans. Amer. Soc. Civil Engrs.*, vol. 53.

Hazen, Allen. 1907. Clean Water and How to Get It, John Wiley & Sons, New York, N.Y.

Hering, Rudolph. 1881. "Sewerage Systems," *Trans. Amer. Soc. Civil Engrs.*, vol. 10.

Jervis, J.B. 1876. "A Memoir of American Engineering," Trans. Amer. Soc. Civil Engrs., vol. 5, pp. 39-67.

Massachusetts State Board of Health. 1890. *Twenty-First Annual Report of the State Board of Health.*

Massachusetts State Board of Health. 1874. *Fifth Annual Report of the State Board of Health.*

Massachusetts State Board of Health. 1894. *Twenty-Fifth Annual Report of the State Board of Health.*

Massachusetts State Board of Health. 1903. *Thirty-Fourth Annual Report of the State Board of Health.*

McCracken, R.A. and D. Sebian. 1988. "Lawrence Experiment Station," *The Diplomate*, vol. 11, no. 2, pp 12-18.

McKinney, A. 1994. "A Stroll Through Time in Environmental Engineering", Environmental Engineer, vol. 31, no. 1, pp 12-13, 25, 33.

Metcalf, L. and H.P. Eddy. 1915. *American Sewerage Practice*, McGraw-Hill, New York, N.Y.

Odell, F.S. 1881. "The Sewerage of Memphis," *Trans. Amer. Soc. Civil Engrs.*, vol. 10, pp 23-52.

Wylie, F.E. 1975. *M.I.T. in Perspective*, Little, Brown & Co., Boston, Mass.

Author's Note

Much of the information in this article has been previously published in *Environmental Engineering P.E. Examination Guide and Handbook, Second Edition*, 2001, Pine Tree Press, Annapolis, Maryland.

A BRIEF HISTORY OF AERATION OF WASTEWATER

WILLIAM C. BOYLE[1], PhD, P.E., DEE

Diffused Aeration

Diffused aeration is defined as the injection of air or oxygen enriched air under pressure below a liquid surface. Although the aeration of wastewater began in England as early as 1882 (Martin,1927), major advances in aeration technology awaited the development of the activated sludge process by Arden and Lockett in 1914. A review of the history of aeration technology is most interesting and instructive. Early investigators were aware of the importance of bubble size, diffuser placement, tank circulation and gas flow rate on oxygen transfer efficiency. Early aeration was provided by perforated tubes and pipes. One of the earliest patents for a diffuser was in 1904 in Great Britain for a perforated metal plate diffuser (Martin, 1927). In Great Britain porous tubes, perforated pipes, double perforated tubes with fibrous material in the annular space and nozzles were tried early on (Federation of Sewage and Industrial Wastes Associations, 1952). Investigators sought more efficient aeration through the development of finer bubbles. In England, experiments were conducted with sandstone, firebrick, mixtures of sand and glass, and pumice. Most of these early materials were dense, creating high head losses. A secret process employing concrete was used to cast porous plates that were placed in cast iron boxes by Jones and Atwood, Ltd. in about 1914, a system that was used for many years by Great Britain and its colonies.

Meanwhile, in the US, porous plates produced by Filtros were widely used in newly constructed activated sludge plants. At Milwaukee, research was conducted using grids of

[1]Emeritus Professor, Civil & Environmental Engineering, University of Wisconsin, Madison, WI 53706

perforated black iron pipes, basswood plates, Filros plates and air jets. The Filtros plates were selected for the plant placed in operation in 1925 (Ernest, 1994). The Filtros plates patented in 1914 were constructed from bonded silica sand and had permeabilities (see Section 3.4.1) in the range of 14.1 to 20.4 m^3_N/h (9 to 13 scfm) at 5 cm (2 in) water gage. Similar plates were installed in the Houston North-Side plant in 1917, as well as at Indianapolis, Chicago, Pasadena, CA, Lodi, CA and Gastonia, NC (Babbitt, 1925). Ernest (1994) provides an excellent history of the development of the aeration system at Milwaukee where siliceous plates from Ferro Corp (Filtros) are still used. Over time aluminum oxide bonded with a variety of bonding agents as well as silica were the major media of choice. Permeabilities continued to rise as well, up to as high as 188 m^3_N/h (120 scfm). In addition, new shapes were introduced including domes and tubes, and, more recently discs.

In Great Britain, the sand-cement plates were predominately used up to about 1932. In 1932, Norton introduced porous plates bolted at either end. Norton introduced the first domes in 1946 with permeabilities in the range of 62.8 to 78.5 m^3_N/h (40 to 50 scfm). In Germany early aeration designs (commencing about 1929) incorporated the Brandol plate diffusers produced by Schumacher Fabrik. Later they developed a tube design and the material was modified as silica sand bonded by a phenol formaldehyde resin (Schmidt-Holthausen and Bievers, 1980).

Diffuser configuration was considered to be an important factor in activated sludge performance even as early as 1915. The Houston and Milwaukee plants were designed with a ridge and furrow configuration. In 1923 Hurd proposed the "circulatory flow" or spiral roll configuration for the Indianapolis plant. The Chicago North-Side plant also employed this diffuser configuration (Hurd, 1923). The design was promoted on the belief that the spiral roll would provide a longer contact time between wastewater and air than the full floor coverage. One set of basins at Milwaukee were converted to spiral roll in 1933 but even the 1935 data base suggested that the spiral roll configuration required more air per unit volume of wastewater treated. The spiral roll configuration was abandoned at Milwaukee in 1961 after extensive oxygen transfer studies (Ernest, 1994). It is also interesting to note that the early plants employed a range of diffuser densities (percent of floor surface area covered by diffusers, A_d/A_t x 100) ranging from about 25 % at Milwaukee and Lodi, CA to 7 to 10 % at the spiral roll plants (Babbitt, 1925).

Clogging of diffusers appears to have been a problem in some cases from the earliest studies. Generally speaking the porous diffusers produced the greatest concern but examples of clogging of perforated pipes can be found (Martin, 1927, Ernest, 1994). Early work by Bushee and Zack (1924) at the Sanitary District of Chicago prompted the use of coarser media to avoid fouling. Later, Roe (1934) outlined in detail numerous diffuser clogging causes. Ernest (1994) detailed cleaning methods adopted by Milwaukee in maintaining porous diffusers at their installations. Nonetheless, by the 1950's many plants were using the large orifice type of diffuser. The newer designs improved upon their earlier counterparts and were designed for easy maintenance and accessibility. In general, these devices produced a coarser bubble, thereby sacrificing substantial transfer efficiency. The 1950 Air Diffusion in Sewage Works manual (Federation of Sewage and Industrial Wastes Associations,1952) provides an excellent summary of air diffusion devices proposed and tested between 1893 and 1950. It should be emphasized that the trend toward coarser diffuser media was followed in the US but not in Europe where the porous diffusers continued to predominate in many designs.

With the emphasis on more energy-efficient aeration in the 1970's in the US, porous diffuser technology received greater attention. Since about 1970, the wastewater treatment industry has witnessed the introduction of a wide variety of new diffuser materials and designs. Many of the lessons learned with this technology in the early part of the century were revisited. Improvements in materials of construction, blower designs and measurement technology has resulted in a new generation of highly efficient diffuser systems and the methodologies for maintenance of these systems.

Mechanical Aeration

An alternative to the diffused aeration systems was the mechanical aeration designs, which had been introduced, in the early 1900's. These, too, began to replace some of the older diffused aeration systems where fouling was considered to be a problem. Mechanical aeration is defined as the transfer of oxygen to water by mechanical devices so as to cause entrainment of atmospheric oxygen into the bulk liquid by surface agitation and mixing. In addition, equipment that causes dispersion or aspiration of compressed air, high purity oxygen, or atmospheric air by shearing and pumping action by a rotating turbine or propeller may also be included.

It appears that mechanical aeration in wastewater was introduced to overcome problems with diffuser clogging in activated sludge systems. The concept was introduced in Europe in the late 1910's, predominantly in the UK, and spread to the US slowly. By 1929, mechanical aeration plants outnumbered diffused aeration plants in the UK by 2:1. In the US, a survey by Roe (1938) indicated that about 100 activated sludge plants employed mechanical aeration, 200 were using diffused aeration, and approximately 20 had combined aeration systems.

Porous tile diffuser clogging in Sheffield, England spurred the development of an Archimedian screw-type aerator in 1916. In 1920, Sheffield built a full-scale facility using submerged horizontal paddle wheels in narrow channels (1.2 to 1.8 m) [4 to 6 ft] that were about 1.2m (4 ft) deep, called the Haworth System. Located midway between the channel ends that interconnected each aeration tank, the shaft rotated at 15 to 16 rpm producing a longitudinal velocity of 0.53 m/s (1.75 ft/sec). The movement of wastewater along the channel created a wave action that allowed transport of oxygen from atmospheric air to the water. The use of pumps to replace the paddles in moving wastewater along the channels did not provide sufficient oxygen transfer and were supplemented by submerged paddles in order to satisfy oxygen demand. Triangular paddles which replaced the rectangular paddles in 1948 improved performance by 40 to 50% when the shaft was operated at twice the original rotational speed.

The Hartley aeration system was similar to that used at Sheffield but employed propellers fixed to inclined shafts. These units were located at the U-shaped ends of the shallow interconnected channels. A series of diagonal baffles were located at intervals along the channels, set at an angle in the direction of flow, to reduce the velocity, to prevent suspended solids separation, and to create new liquid surfaces to come in contact with the atmosphere. These systems were used at Birmingham and Stoke-on Trent in the UK. Neither the Haworth nor the Hartley system found service in the US.

Other horizontal shaft systems were also being developed in these early years. In 1929, pilot studies at Des Plaines, IL were described in which an aeration device employing a steel latticework attached to a horizontal shaft to form a paddlewheel. The paddlewheel, with a diameter of 66 to 76 cm (26 to 30 in), was suspended along the entire length of the aeration tank and was partially submerged so that when rotated it would agitate the liquid surface. A vertical baffle, running along the entire basin length 46 cm (18 in) from the wall and located below the paddlewheel, terminated at the surface with a narrow trough located at right- angles to the basin wall. The shaft was rotated in the range of 36 to 60 rpm by an electric motor. This rotation toward the wall caused mixed liquor to rise upward between the baffle and wall and fall downward in the main basin. The wave-like motion at the surface created new liquid surfaces to contact the air. Mixed liquor flowed in a spiral roll configuration down the aeration tank. This system was known as the Link-Belt aerator. Link-Belt aerators were installed in several US plants in the 1930s, but were not in production by the late 1940's.

Another horizontal rotor device often referred to as a brush aerator was developed in the US and Europe in the 1930's. Called brush aerators because of the use of street cleaning brushes during early development, these devices were usually fastened to one longitudinal wall of the aeration tank and partially submerged below the liquid surface. Rotating at speeds ranging from 43 to 84 rpm, the brushes created a wave-like motion across the liquid surface and induced a spiral roll to the wastewater as it flowed down the aeration tank. Kessner employed brushes as well as a combination of brushes and submerged paddles in Holland as early as 1928. Similar in design and function as the brush aerators described above, Kessner employed the submerged paddles mounted on a horizontal shaft and rotated at 3 to 7 rpm to supplement the brush and to provide a reinforced spiral roll to the mixed liquor. The newer Kessner brushes employed acute triangles cut from stainless steel sheet in place of the brush. The aeration tank bottom was either rounded or the sidewalls sloped near the bottom to enhance circulation.

An interesting modification of the horizontal paddle aeration system resulted in the combination of paddlewheels and diffused air developed in Germany and reported by Imhoff in 1926. The submerged paddles, made of steel angles and mounted on horizontal shafts running longitudinally along the aeration basin, were rotated counter to the upward flow of bubbles. Diffused air was provided longitudinally along the wall or center- line of the tank.

In addition to the horizontal rotor concept, vertical draft tube aerators were also being developed at this time. In the early 1920's the Simplex system was developed in the UK. The Simplex system in its earliest version employed at Bury, England was a vertical draft tube device placed in a relatively deep hopper-bottom tank. A vertical steel draft tube with open bottom located about 15 cm (6 in) from the floor was suspended at the tank center. At the top of the tube was a cone with steel vanes. The cone was rotated at about 60 rpm drawing mixed liquor up through the draft tube. The wastewater was then sprayed outward over the surface of the tank. Each draft tube was driven by its own motor through a speed reducer or by a line shaft with individual clutches. A number of vertical draft tube systems resembling the Simplex aerator became popular in the US in the 1930's and 1940's. They were the predominant mechanical aerators in the US by the 1950's.

In the 1950's and 1960's many low-speed aerators were sold in the US and they apparently performed satisfactorily. However, there was no generally accepted method of evaluation of the units and the main testing efforts were aimed at process performance. A major flaw soon became apparent, however, relative to aerator maintenance and reliability- the gear reducers. Often gear reducers failed within a short period of time after initial start-up. Some lasted for a year or two, but many failed after only a few weeks or months.

From a performance perspective, the 1950"s vintage impellers were almost all simple radial flow devices. A number of impeller designs were imported from Europe and adopted by US suppliers. Innovative blade designs were developed as well by US manufacturers. At that time and into the 1960's, no reliable test procedure was available to assess the value of these designs. The effects of impeller speed, basin geometry and other important dependent variables were either unknown or poorly understood. As a result, it is likely that most systems were under designed. On the other hand, virtually all worked to the satisfaction of the operators with the exception of mechanical problems.

By the mid 1970's, the mechanical problems had been recognized and, at least to a certain extent, addressed by the major aerator suppliers. The dynamics of the market were changing at that time, with the old-line equipment suppliers being squeezed by newer entrants. Since the 1960's Lightnin, a major manufacturer of mixers, made a big push in the low speed aerator market using a very inexpensive impeller (a 4-blade pitched blade turbine). Shortly thereafter, Philadelphia Gear's Mixing Division entered the market using specially designed reducers and new impellers that were less prone to cause failures. From a mechanical perspective, these new suppliers represented the best level of quality ever seen in the business at a cost that the older manufacturers found hard to match. At least as important, these mixing companies were very familiar with the best approach to blending liquids and suspending solids. As a result, by the mid 1980's, the leading low-speed aerator manufacturers in the US were Lightnin and Philadelphia Mixers. That situation still exists as we enter the 21st century since no new low-speed aerator suppliers have come into the market in the last 30 years.

Today, the low-speed surface aerator remains a very popular device in certain niches. High-purity oxygen suppliers have found that good low-speed aerators do the best job for their process and Eimco continues to be successful in their Carrousel™ ditch process using the low-speed vertical shaft machines. In addition, many low-speed units are performing well in activated sludge systems.

For lagoon applications and situations where capital cost is a major factor, several manufacturers began to offer motor speed or high-speed aerators in the 1970's. Primarily of a floating configuration, the development aimed at lagoons and small extended aeration facilities. All used marine propellers as the impeller of the non-snagging type. In the early days of development, these devices were plagued by mechanical difficulties largely due to motor bearing failures as well as poor manufacturing quality control. The hydraulic forces were the main cause of bearing failure and it took a while for manufacturers to find effective designs to ensure long-term service. New styles of motor speed devices are currently being designed and marketed. Because of their popularity, innovation continues to improve performance and reliability.

At the same time the low-speed aerator was being improved in the 1960's, the horizontal rotor became popular in oxidation ditch applications in the US and in Europe. A number of different rotor designs have been used, ranging from brushes to the more complex discs. Their efficiency is consistent with the radial flow style of low-speed aerator impellers, and similar concerns regarding mechanical integrity (gear reducer and bearings) have been addressed and largely overcome.

Also designed for lagoon applications, aspirating devices became popular in the US in the 1970's. A number of different configurations have been used, which includes a floating device that uses a marine propeller mounted at a shallow angle to the horizontal and a submersible pump unit using a vertical draft tube. Fashioned in a way that allows air to be aspirated through its hollow shaft, these devices are effective mixers, adding some oxygen in the process. These units have also experienced a series of historical mechanical difficulties, mainly associated with shaft-supported bearings located below the water surface. A number of approaches have been used in an effort to resolve these problems. The problems remain and, although very inexpensive, they do not provide top performance or trouble free operation

Finally, in this brief historical overview, are the submerged, sparged turbine aerators that have been used for decades in a number of forms. Industrial mixing requirements often have called for the introduction of a gas into a liquid. The major mixing companies in the US (Lightnin, Philadelphia Mixers , and Chemineer) were all familiar with the concept. In the 1960's and 1970's, several companies tried to improve the surface aerator performance by designing aerators that would disperse compressed air using what is essentially a mechanical mixer. Two general types were developed at that time- the radial and draft tube (radial) and an open-style axial flow type (down-pumping impeller above the sparger). These units were plagued with mechanical problems and did not perform as well as anticipated. As a result, they have fallen out of favor in today's market. The draft tube turbine aerator is similar in concept in that it uses a down-pumping impeller positioned above an air-release device. The impeller and sparger are located within a draft tube that assists flow direction and shearing action. These devices, used in deep basins (7.6 to 9.75 m)[25 to 32 ft), have experienced some early mechanical failures that have recently been overcome. A radial flow submerged turbine aerator uses a radial flow impeller positioned above a sparger. Offered in the early 1960's and still used today in aerobic digestion applications, its mechanical reliability is high.

References

Babbitt, H.E.(1925). *Sewerage and sewage treatment,* Second Edition, John Wiley and Sons, New York.

Bushee, R.G. and Zack, S.I. (1924). "Tests of air pressure losses in activated sludge plants." *Engineering News Record,* 93, 823.

Committee on Sewage and Industrial Waste Practice (1952). *Air diffusion in sewage works MOP5,* Federation of Sewage and Industrial Waste Associations, Champaign, IL.

Ernest, L.A. (1994). *Case history report on Milwaukee ceramic plate aeration facilities,* EPA 600/R-94/106, NTIS No. PB94-200946, EPA, Cincinnati, OH.

Hurd, C.H. (1923). "Design features of the Indianapolis activated sludge plant", *Engineering News Record,* 91, 259.

Martin, A.J. (1927). *The activated sludge process,* MacDonald and Evans Publ., London, England.

Roe, F.C. (1934). " The installation and servicing of air diffuser mediums", *Water Works and Sewerage,* 81, 115.

Roe, F.C. (1938). "The activated sludge process- a case for diffused aeration", *Sewage Works Journal,* 10, 999.

Schmidt-Holthausen, H.J. and Zievers, E.C. (1980). "50 years of experience in Europe with fine bubble aeration", *53rd Annual Conference-WPCF,* Las Vegas, NV.

The Metropolitan Water Reclamation District of Greater Chicago; Our Second Century of Meeting Challenges and Achieving Success

Joseph T. Zurad[1], Joseph P. Sobanski[2], Joseph R. Rakoczy[3]
and Richard Lanyon[4]

Abstract

Sited on a marshy, level lake plain with small, sluggish natural rivers and streams, throughout its existence Chicago (chartered 1837) has been faced with a multitude of drainage and water-related public health challenges ranging from mid to late 19th century typhoid, cholera, and dysentery epidemics, to latter day combined sewer overflow waterway pollution, overbank and sewer backup flooding, and lake beach closings due to contaminated river backups induced by severe rainstorms. These problems stem from Chicago's tremendous growth, which to varying degrees has continually strained and sometimes breached the natural and manmade water/wastewater infrastructure. The Sanitary District of Chicago (in 1989 renamed the Metropolitan Water Reclamation District of Greater Chicago) was created in 1889 to restore and maintain water-related public health by protecting Lake Michigan – the area's main drinking water supply – from sewage contamination. It did so in a pioneering way that captured the attention of the world, contributed to a Chicago reputation later verbalized as a positive "make no small plans" attitude, significantly advanced civil engineering and civil works construction technology and techniques, and helped spur other major earth moving projects elsewhere. As Chicago urbanized and industrialized further, and as environmental science and conscience heightened so as to recognize the need for higher water quality standards, new water management challenges arose which the District addressed, and continues to address, by undertaking a succession of other notable civil works efforts.

----

[1]Chief Engineer, [2]Assistant Chief Engineer, [3]Supervising Civil Engineer and [4]Director of Research and Development; Metropolitan Water Reclamation District of Greater Chicago, 100 E. Erie Street, Chicago, IL 60611; phone 312-751-5600; e-mails: joseph.zurad/joseph.sobanski/joseph.rakoczy/dick.lanyon@mwrd.org

Introduction

The District today is an independent municipal government and taxing body headed by a nine-member, elected Board of Commissioners. Its 2,258 km^2 (872 sq. mile) expanded service area covers 91% of the area of Cook County, Illinois, and includes the City of Chicago and 124 suburban cities and villages. Population equivalent served is 5.1 million people and 4.5 million commercial/industrial equivalent. The District's key responsibilities are to protect Lake Michigan from pollution and to collect and treat wastewater to prevent inland waterway pollution. The District owns and operates seven water reclamation plants having a total design treatment capacity of 9.46x10^6 m^3/day (2.5 billion gal/day). The 5.4x10^6 m^3/day (1.44 billion gal/day) Stickney Water Reclamation Plant is one of the world's largest. The District also has 892 km (554 miles) of intercepting sewers, over 161 km (100 miles) of combined sewer overflow control deep tunnels, 23 pumping stations, and 31 flood control reservoirs. It also maintains 132 km (82 miles) of navigable urban waterways that serve as the mid-continental water connection linking the north Atlantic Ocean (Canada) and the Gulf of Mexico.

A number of the District's civil works efforts have been referred to as "generational," with major pollution and flood control infrastructure added to the water management system as Chicago's growth resulted in new or worsening drainage and sewage disposal problems. Other efforts have supplemented the system, and while they may have less overall impact, they nevertheless were necessary, innovative and have won acclaim in their own right.

The First Act - Reversal of the Chicago River

First drawn to the region by the promise of water commerce afforded by favorably positioned and slow moving rivers, and by Lake Michigan's abundant supply of fresh drinking water, the people of Chicago have always been aware of the vital importance of their water resources. Its poor natural drainage coupled with booming city growth particularly in the two decades after the 1871 Great Chicago Fire, eventually led to the disastrous situation of residents drinking the city's collective sewage. The city's sewers captured street runoff, human and industrial waste - including stockyard and slaughterhouse wastes – and discharged them to the river, which flowed into the lake. As the city continued to grow, lake water intakes were placed further off-shore to reach clean water, but in 1885 a huge rainstorm flushed the river and its sewage far into the lake, beyond the water intakes. Ensuing disease epidemics in the mid-1880s killed thousands of the city's 750,000 residents, earned Chicago the temporary nickname "Death City," and raised a public outcry for a permanent solution to the health crisis.

In 1889, the State of Illinois enabled the creation of the Sanitary District of Chicago (later called the Metropolitan Sanitary District of Greater Chicago, and since 1989 the Metropolitan Water Reclamation District of Greater Chicago) which was charged with achieving one goal – safeguard Lake Michigan from sewage contamination. The civic push for the District's formation in hopes of ending the health crisis was astounding; the referendum vote for establishing the District - and increasing the taxpayers' taxes - was 70,958 for, 242 against.

District engineers solved the problem in a bold and ambitious way – by reversing Chicago's polluted rivers so they flowed backwards, away from the lake. This represented the first time a river was redirected to flow away from its mouth. A system of three canals totaling 113 km (70.5 miles) in length was built from 1892 to 1922, first with the pioneering 45 km (28 mile) long, 7.3 m (24 ft.) deep, and 48.8 m (160 ft) wide Sanitary and Ship Canal begun in 1892 and placed in operation in 1900, and later with the North Shore Channel (1910) and the Cal-Sag Channel (1922). In a monumental effort involving 8,500 workers, mules, picks, shovels, steam-powered machines, explosives, and innovative excavation techniques, the Sanitary and Ship Canal was cut through a low point on the sub-continental divide located a few miles west of Chicago separating the east flowing Chicago River and the Great Lakes basin from the southwest flowing Des Plaines River and the Mississippi River basin. The canals diverted the polluted Chicago rivers to flow to the Mississippi River and the Gulf of Mexico and away from the Lake Michigan drinking water supply (see figure 1).

The Sanitary and Ship Canal cost about $70,000,000 to build. Once completed, waterborne disease rates quickly dropped, and Chicago, which had the highest typhoid disease rate of all major cities in the U.S., Canada and Europe, was soon regarded as being one of the safest in the world. With its water supply protected by the canals, Chicago and the region grew and prospered.

The reversal of the Chicago River system was one of the largest municipal public works projects ever completed, and was hailed as one of the greatest engineering achievements of all time. Widely referred to as the "eighth wonder of the world," in 1955 the American Society of Civil Engineers named it and the District "One of the Seven Engineering Wonders of American Engineering" in recognition of the river's reversal and the vast wastewater collection and treatment system. In 1999, *Engineering News Record* named the project "One of the Top 125 Construction Projects of the Past 125 Years."

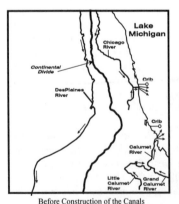

Before Construction of the Canals   After Construction of the Canals

Figure 1. The Reversal of the Chicago and Calumet Rivers.

Significant new excavation techniques developed on the project collectively came to be known as the "Chicago School of Earth Moving" and were instrumental in the feasibility and construction of the Panama Canal. Now, at the start of the 21st Century, the canals continue to serve crucial pollution control, flood prevention, and navigational needs.

Locks and Dams

River reversal meant polluted river flows were being flushed and diluted by Lake Michigan water pouring into the inland waterways at three points - the only Great Lakes outflow points other than the Saint Lawrence River natural outlet. Fears were raised among neighboring Great Lakes states and Canada that Chicago's project would adversely draw down lake levels, affecting water availability and impacting fisheries and lake shipping. The channels were designed for 283.2 m^3/sec (10,000 ft^3/sec) flow, but after litigation, the U. S. Supreme Court imposed strict total limits on the amount of lake water that could be diverted for runoff, pollutant dilution, leakage, lockage, and navigational make-up. To comply with the court decree, help protect the lake from river backflows, and maintain navigation between the inland waterways and the lake, the District built flow-controlling locks and dams on the waterway intakes. These facilities – after improvements over the years – exist today as the Chicago River Controlling Works, at the mouth of the Chicago River near downtown Chicago; the O'Brien Lock and Dam, on the Calumet River south of Lake Calumet; and Wilmette Harbor controlling works, on the lake at the mouth of the North Shore Channel. The 183 m (600 feet) long and 24.3 m (80 feet) wide Chicago River lock passes more than 12,000 boats each year. The Lockport Lock and Dam, at the downstream end of the District waterway system, includes a hydropower station producing electricity fed into the local power company's grid, saving the District about $3 million annually. In 1984, operation and maintenance of the locks at the mouth of the Chicago River was turned over to the United States Army Corps of Engineers.

Built in 1991, the Waterways Control Center, in the District's main administration building in downtown Chicago, is a state-of-the-art control room and operations hub for the 132 km (82 mi) waterway system. The Center is manned 24 hours a day and receives telemetered data from 23 automatic rain gauges, numerous canal, deep tunnel and lake level gauges, and real-time and predictive weather radar. The computerized data is used in real-time decision making to best maintain waterway navigation depths, maximize available system storage capacity to prevent excessive flood stages, regulate flow to meet diversion limits, and protect Lake Michigan from polluted backflows during severe storm events.

Sewage Treatment Plants

The District was one of the first agencies in the nation to design and build large-scale sewage treatment plants, and it has been a leader in the development and implementation of secondary and advanced wastewater treatment technology.

After the reversals of the Chicago and Calumet Rivers in the early 1900s, St. Louis and other downstream communities contended that Chicago's river-borne wastes created a health threat to their water supplies. Lake water diversion was used to facilitate water cleansing in a then conventional treatment approach referred to as "dilution is the solution to pollution." A federally monitored field test using a bacterial indicator organism released in the waterway system at Chicago showed there was no pollution threat to St. Louis. However, Chicago's continued growth, along with the Supreme Court diversion restrictions, quickly outstripped the natural self-cleansing ability of the overall river system system. In the 1920s, the District built two treatment plants, among the first modern large-scale, activated sludge plants in the nation (Calumet and North Side). The Stickney sewage treatment works (today, after expansions, the Stickney Water Reclamation Plant) was built beginning in the 1930s and is one of the largest treatment plants in the world. With an average daily flow of almost 3.0×10^6 (800 million gal) and a total design capacity of 5.4×10^6 m^3/day (1.44 billion gal/day), the plant now covers over 202 ha (500 acres) and serves more than 2 million people. Today the District has seven water reclamation plants with a total design capacity of 7.6×10^6 m^3/day (2.5 billion gal/day).

After sewage processing, plant effluents meeting environmental water quality standards are discharged to local waterways flowing away from Lake Michigan. The District has won numerous awards for the operating performance of its plants and compliance with effluent water quality standards. According to a 1999 survey by the American Municipal Sewage Agencies (AMSA), among 119 of the nation's wastewater treatment agencies with populations greater than 1.5 million, the District achieved the lowest average cost for collection and treatment of wastewater. In 2000, all seven District plants received AMSA Gold Awards for 100% prior-year compliance with NPDES water quality permit standards. In 1999, an AMSA survey of 119 member agencies revealed the District had the lowest average cost for collection and treatment of wastewater ($408/MG).

Intercepting Sewers

As part of its treatment plant construction program begun in the 1920s, the District built large trunk sewers along the rivers and canals to intercept and divert wastewater flows away from the rivers and to the new treatment plants. The District now owns and operates 891 km (554 mi) of such interceptors and force mains ranging from 30.5 cm to 8.84 m (12 inches to 29 feet) in diameter. Thousands of miles of local sewers owned by 125 villages and cities, including the city of Chicago, tie to the intercepting sewers at an estimated 2,200 connection points. Early sewers were built of concrete, tile, and brick and were excavated by various open cut and mining methods. Today, interceptors and relief sewers are often installed using more modern methods including soft and hard ground tunnel boring machine mining. Interceptors are designed to provide capacity for the anticipated peak ultimate population and industrial flow. Since virtually all of the District's area is serviced by existing interceptors, the District's interceptor programs now consist mainly of maintenance and rehabilitation of existing facilities.

The Prairie Plan/Biosolids Disposal

Over 181,000 MT (200,000 tons) of residual biosolids result from the District's sewage treatment processes annually and must be disposed. The District has been a pioneer in developing beneficial ways to use these biosolids, with many Chicago area golf courses, sod farms, highway banks, and parks benefiting from the successful processing and application of this material.

In the early 1970s, the District undertook an ambitious plan to use biosolids to restore over 6,070 ha (15,000 acres) of strip-mined land in Fulton County, about 322 km (200 mi) southwest of Chicago. Called the "Prairie Plan," the project merged its biosolids disposal needs with local land reclamation desires. County officials wanted to transform the low viability, strip-mined area to land suitable for farming and recreation. The Prairie Plan project turned the scarred, infertile land into rich farmland producing crops of corn, wheat and soybeans not possible before the application of the biosolids. Conservation areas with nature trails and camping facilities have been integrated into the project. To date, over one million dry tons of bio-residuals have been applied to the site. The project won the United States Environmental Protection Agency Beneficial Use of Sludge Competition and a number of other awards including the ASCE 1974 Outstanding Civil Engineering Achievement of the Year Award.

Sidestream Elevated Pool Aeration (SEPA) Stations

In the late 1970s, the District undertook an innovative and attractive method of aerating oxygen-starved waterways. Two instream aeration stations were built on the North Branch to supplement natural reaeration. In the 1990s, five "sidestream elevated pool aeration (SEPA)" stations were built at strategic locations on the Calumet River and the Cal-Sag Channel in Chicago and three southern suburbs. The SEPA stations, also called "Urban Waterfalls," replenish dissolved oxygen in the waterways by replicating one of nature's simplest water treatment processes. Utilizing 2.74 m (9 ft) diameter, 12.2 m (40 ft) long Archimedes screw pumps, up to half the river flow is diverted into the stations, lifted 3.6 m to 4.6 m (12 to 15 feet, and allowed to cascade in a series of pools and drops back to the waterway. Oxygenation results in the lifting, pooling, and especially the waterfalling stages. The oxygen added to the river allows water quality standards for dissolved oxygen to be met and has spurred a dramatic return of fish and other aquatic wildlife populations. The stations were designed to complement the surrounding areas and are attractively landscaped with park-like settings. In praising the project, the *Chicago Tribune* cited the waterfalls as good places for picnics, relaxation, and even wedding photography.

The ASCE selected SEPA for its 1994 Outstanding Civil Engineering Achievement Award, commenting "Deceptively simple in concept, Chicago's SEPA system meets multiple engineering challenges: substantial water quality improvement; low cost and high efficiency; preservation of wildlife and navigation; and value to surrounding communities. The first major use of artificial waterfalls for pollution control, SEPA is a model of creative civil engineering solutions to complex urban problems."

Tunnel and Reservoir Plan (TARP, or "Deep Tunnel")

In the mid-1960s, metropolitan Chicago began searching for a comprehensive solution to its massive and worsening combined sewer overflow (CSO) pollution and flooding problems. Within a 971 km^2 (375 sq mi) combined sewer area with over 3 million people in 52 cities and villages, including the city of Chicago, rainfalls as small as 1/3 inch overloaded local sewer systems and caused CSOs – a mixture of raw sewage and stormwater runoff – to spill to area rivers from 438 local and 38 District sewer outfalls. The CSOs severely polluted area waterways, and, with the heaviest storms, raised flood stages to levels resulting in river backflows to Lake Michigan, closing beaches. A main underlying cause of the problem was lack of an adequate floodwater outlet in the face of continued urbanization.

Numerous studies and the cooperation of many agencies led to the Tunnel and Reservoir Plan (TARP, or Deep Tunnel) - a hybrid of the best elements of over 50 alternatives examined – being selected as the one solution cost-effectively fulfilling state and federal CSO water quality requirements and meeting three water management goals: protect Lake Michigan from river backflows; clean up area waterways; and reduce street and basement sewage backup flooding.

TARP was officially adopted by the District in 1972 and started under construction in 1975. The large scale, multi-purpose TARP concept – huge, deep rock tunnels and reservoirs that capture, convey and store combined sewage until transferred to existing treatment plants when capacity becomes available following storms – was at first controversial due to its seemingly large cost and unproven nature. However, successful performance of major tunnel segments placed on line beginning in 1985 and the attainment of many significant side benefits, have made TARP a worldwide model urban water management tool.

Phase I of TARP – intended primarily for pollution control - was the nation's first comprehensive CSO control plan for a large urban area. Designed to capture approximately 85% of CSO first-flush pollution, Phase I entails 176 km (109.4 mi) of deep tunnels ranging from 2.7 to 10.0 m (9 to 33 feet) in diameter and from 45.7 m (150 ft) to more than 91.4 m (300 ft) deep; over 250 drop shafts; three pump stations; and over 600 surface connecting and flow control structures.

Phase II is mainly for flood control but will also yield significant pollution control benefits. Consisting of three large, surface reservoirs, the current Phase II plan is called the Chicagoland Underflow Plan (CUP) and is being designed and constructed by the U. S. Army Corps of Engineers. The 1.3x10^6 m^3 (350 million gal) O'Hare Reservoir was placed in operation in 1998. Constructed in stages, the McCook Reservoir will have a capacity of 40x10^6 m^3 (10.5 billion gal), and the Thornton Reservoir will provide 18x10^6 m^3 (4.8 billion gal) of storage.

Presently, 150.3 km (93.4 mi) of tunnels are on line, with the remaining 25.7 km (16.0 mi) under construction to be completed by 2006.

TARP's performance thus far has been remarkable by many measures. To date, the project has captured and allowed treatment of over 2.42x10^9 m^3 (640 billion gal) of CSO - a volume equal to a continuous line of tank trucks circling the earth nearly 4 times. Over 2.81×10^8 kg (620 million lbs) of oxygen consuming substances (BOD_5 plus ammonia) have been captured.

Dramatic river water quality improvements have led to three-fold and higher increases in fish populations, with over 50 species of fish, including game fish, now inhabiting local urban waterways. Marked visible improvements to the waterways have led to booming increases in riverside land values and development including hotels, office/condominium buildings, restaurants, riverwalks, canoe launches and marinas, and has stimulated vigorous recreational uses of the Chicago River, including tourism, boating, and fishing. The year 2000 Bassmaster fishing tournament was held in Chicago; first for a large city.

Cutting edge advancements in tunneling technology and techniques have been developed on TARP, including major improvements in tunnel boring machine design and use. Numerous mining rate world-records have been set on TARP, and tunneling efficiencies have greatly lowered overall construction costs and spurred other tunnel projects such as the English Channel Tunnel (Chunnel).

TARP Phase I and Phase II/CUP is estimated to cost $3.1 billion. To date, $2.2 billion in TARP Phase I projects have been completed and placed in service, and $230 million remain. Remaining Phase II/CUP reservoirs now under final design and in the preliminary stages of construction will cost $658 million.

Over $1.6 billion (1985 estimate) in treatment plant expansion savings and $692 million in new interceptor and local relief sewer savings have been realized due to TARP's detention storage which reduces peak flows and optimizes existing sewer and treatment capacity. Significant side benefits and uses of TARP, some unexpected, include use as a bypass conduit for emergency interceptor and pump station repairs and for planned rehabilitation of interceptors and as an outlet during rising canal levels to prevent river backflows to the lake.

TARP has received much recognition and numerous awards from government and professional engineering and environmental organizations. In 2000, *Engineering News Record* named TARP "One of the Top 125 Construction Projects of the Past 125 Years." ASCE awarded TARP Phase I the 1985 Outstanding Civil Engineering Achievement. In 1997, the United States Environmental Protection Agency (USEPA) cited TARP in declaring the water quality restoration of the Chicago River and Illinois River as one of the top Clean Water Act (CWA) Success Stories.

Water Quality Monitoring

Since the early 1970s, the District has operated an ambient water quality monitoring program to assess the condition of its receiving waters. Today, this program consists of monthly grab sampling at 59 locations and analysis for over 50 physical, chemical and bacterial parameters. The monitoring data is contained in a local database as well as being loaded on the USEPA's STORET database. The program is governed by a Quality Assurance Project Plan approved by the Illinois Environmental Protection Agency (IEPA) and IEPA uses this data in their annual water quality conditions report required by Section 305(b) of the CWA. In addition, the data is used by IEPA to prepare the Section 303(d) list to determine the need for total maximum daily loads. It should be noted that the District operates its own analytical laboratories and these have received accreditation by the IEPA under the National Environmental Laboratory Accreditation Program.

Recently, the monitoring program has been expanded to include assessment of biological conditions, physical habitat, sediment quality and sediment toxicity at the same 59 locations. The frequency of these assessments at each location is either annual or quadrennial, depending upon location. Those locations in proximity to a District WRP are assessed annually. The District is currently in the process of detailing protocols, quality assurance plans and database formats for these assessments. It is intended that the resulting data will be made available to the IEPA for their use in complying with the same CWA requirements as mentioned above.

To assess the impact of District operations on downstream waters, the District has conducted monitoring of the Illinois Waterway between Lockport and Peoria, Illinois since 1984, covering a distance of 214 km (133 mi) along the Chicago Sanitary and Ship Canal, Des Plaines River and Illinois River. Six sampling runs are conducted by boat each year during which samples are collected at 49 locations and analyzed for over 40 physical, chemical and bacterial parameters. Samples of sediment are collected at 14 locations on one run annually and analyzed for 18 chemical parameters.

The District believes that it is unique in the extent of its monitoring of the receiving waters in its jurisdiction and downstream thereof. The monitoring has demonstrated that District operations have no adverse impact on these waters and also demonstrated that significant improvement in water quality conditions has occurred as a result of CWA programs and funding.

NPDES Permits

The District operates seven water reclamation plants (WRPs) governed by NPDES permits. Four of these WRPs (Calumet, Lemont, North Side and Stickney) discharge effluent to Secondary Contact waters and three (Egan, Hanover Park and Kirie) discharge to General Use waters. These are use classifications established by the State of Illinois. The General Use classification meets the water quality goals of the CWA, but the Secondary Contact standards do not protect for body contact recreation or for the propagation of aquatic life. The Secondary Contact classification is specifically intended for urban waters supporting commercial navigation and drainage from combined sewer areas. The USEPA has a long-standing request for Illinois to review this use classification and attempt to modify it to either meet the CWA goal or demonstrate the inability to meet this goal.

NPDES permits for the WRPs discharging to Secondary Contact waters expired in the early 1990s, prior to the issuance of the USEPA 1994 Combined Sewer Overflow Policy. Reissuance of these permits took a long and difficult path because of the need to follow the policy and to recognize TARP as the long term control plan required in the policy. Permit negotiations eventually concluded in early 2002 and reissued permits became effective March 1, 2002. Distinct from the prior permits, which focused on compliance with effluent quality limits, the new permits focus on compliance with CSO requirements and water quality standards.

Use Attainability Analysis (UAA) Study

The IEPA is about to begin the UAA Study of the Chicago Waterway System (CWS) in 2002. The study outcome will determine the future water quality standards and condition of these waterways. The CWS consists of 125.5 km (78 mi) of canals, which serve the Chicago area for two principal purposes, the drainage of urban storm water runoff and treated municipal wastewater effluent and the support of commercial navigation. Other purposes include recreational boating, fishing, streamside recreation and, to a limited extent, aquatic habitat for wildlife.

The waterway segments (see Figure 1) included in the CWS include the deep draft portions of the Cal-Sag Channel, Chicago River, Sanitary and Ship Canal, Little Calumet River and North Shore Channel. Approximately 75 percent of the length are man-made canals where no waterway existed previous to construction and the remainder are natural streams that have been deepened, straightened and/or widened to such an extent that reversion to the natural state is not possible.

The flow of water in the CWS is artificially controlled by hydraulic structures. These structures include the Chicago River Controlling Works, Lockport Powerhouse and Lock, O'Brien Lock and Dam and Wilmette Pumping Station. Control of flow in the CWS is managed by the District, but is subject to regulation under U.S. Supreme Court Decree and Title 33, Code of Federal Regulations (CFR), Parts 207.420 and 207.425. The CFR provides for the maintenance of navigable depths to support commercial navigation.

A U.S. Supreme Court Decree governs the quantity of water from Lake Michigan that is diverted out of the Great Lakes Basin into the Mississippi River Basin by the State of Illinois. Within Illinois, this quantity is subject to regulation by the Illinois Department of Natural Resources, Division of Water Resources (DWR). The DWR issues allocation orders for annual average quantities of diversion. Most of the diversion is allocated to municipalities for domestic consumption. The District has an order that allows it to divert water for improvement of water quality and this is referred to as discretionary diversion. Currently and through 2014, the District allocation is for an annual average of 7.6 m^3/sec (270 ft^3/sec). In 2015, it is scheduled for reduction to an annual average of 2.9 m^3/sec (101 ft^3/sec).

Over 70 percent of the average annual flow in the CWS is treated municipal effluent from the District's WRPs. Other sources of flow in the system include diversion from Lake Michigan, inflow from tributary streams and storm runoff from storm sewers, streets, parking lots, rooftops, etc. In the cold weather season, nearly all the flow in the CWS is from the WRPs because these other sources are contributing minimum or zero flow. In the warm weather season, less than half of the flow is from the WRPs because the other sources contribute significant inflow. All outflow leaves the CWS at Lockport, however, inflow occurs at numerous locations along the length of the waterways.

A UAA Study, at 40 CFR Part 131.3(g), is defined as "...a structured scientific assessment of the factors affecting the attainment of the use which may include physical, chemical, biological, and economic factors...." The use referred to is the actual existing use of the waterway (boating, fishing, etc.) or the Secondary Contact use classification as designated by the State of Illinois. At 40 CFR Part 131.2, USEPA states that "...water quality standards should, *wherever attainable,* provide water quality for the protection and propagation of fish, shellfish and wildlife and for recreation in and on the water and take into consideration their use and value of public water supplies, propagation of fish, shellfish, and wildlife, recreation in and on the water, and agricultural, industrial, and other purposes including navigation." Emphasis added.

The IEPA is approaching the use attainability issue in stages and began by initiating a UAA Study for the Lower Des Plaines River downstream of Lockport. All the flow leaving the CWS passes through this reach, also designated as a Secondary Contact use water. This study began in 1999 and is expected to end in 2002. The conclusion of this study may result in a petition for a change in the Secondary Contact use classification and/or water quality standards. The potential change would affect at least a portion of this reach, a distance of 19.6 km (12.2 mi). Half of this reach is severely channelized and the entire reach is subject to intense commercial navigation. Any change in use classification and/or water quality standards may have an impact on the CWS and the District. As this study is nearing completion, it appears that the downstream portion of this reach may eventually be upgraded to General Use. This is based on findings that current chemical water quality in this portion meets the current General Use standards. However, the sediment quality in this reach does not appear to support an upgrade.

In the upstream portion of this reach, findings thus far do not support an upgrade. Although chemical water quality comes close to meeting the General Use standards, the sediment quality and the instream and riparian habitat do not support a biological community equivalent to what would be found in a General Use waterway. This is primarily due to the channelized conditions throughout the reach and the accumulated sediments in the reach immediately above a lock and dam.

The District will play a vital role in the conduct of this study, not only because the NPDES permits require that the District "...shall be a participant in and support the UAA that is being undertaken..." but also because the outcome will impact the future of the District. The water quality monitoring data collected by the District will be essential in providing a sound scientific basis for analysis of water quality conditions and the impact of additional treatment technologies. The District is also developing an unsteady-state water quality mathematical model for the CWS to simulate the complex hydraulic and water quality phenomena found in this artificially controlled system. The system is subject to flow reversals and surges caused by lock operation and to density currents caused by chemical and thermal fluxes.

Concurrent with the CWS UAA Study, two other studies will be undertaken by the IEPA regarding Secondary Contact waters in the area. These include Lake Calumet, an off-channel lake along the Calumet River, and the Grand Calumet River (GCR). Lake Calumet receives no effluent discharges, but receives drainage from nearby land areas, industrial sites and landfills. It outlets to the Calumet River, which is already classified as a General Use water. The GCR drains from the northwest corner of the State of Indiana and flows approximately 2.4 km (1.5 mi) in

Illinois before discharging into the CWS via the Little Calumet River. Two Indiana communities discharge treated municipal effluent just upstream of the state line. Numerous existing and former industrial sites are also tributary to the GCR in both states. Since the GCR is an interstate water, coordination with the Indiana Department of Environmental Management is essential.

More Challenges Ahead

In recognition of the District's aging infrastructure, programs are in place to provide the necessary facility maintenance. Extensive rehabilitation is underway of the District's network of 891 km (554 mi) of intercepting sewers, most of which have over 50 years of service. The WRPs will be investigated for rehabilitation or replacement of treatment unit processes. Expansion of treatment capacity is being planned to provide for eventual population and business growth in the service areas for each WRP and to accommodate eventual standards for nutrients in receiving waters.

Design of the TARP reservoirs is being handled by the USACE with involvement of the District's engineers and scientists. These will be formidable in size and provide short-term storage for CSOs until the WRPs can process the contents for treatment. Critical reservoir design elements include the need for aeration to avert odor emissions, controlling the deposition and removal of sediments on the reservoir bottom, periodic cleaning of reservoir sidewalls, loadings imposed on the WRP treatment unit processes and minimizing the energy costs associated with pumping for reservoir dewatering.

Closing

On September 19, 2001, the District was awarded an ASCE Civil Engineering Monument of the Millenium Award for its civil engineering efforts in improving society's quality of life with respect to urban wastewater management.

In executing its projects, the District has had a number of advantages. A regional government with a well-defined mission in water management, it has regulatory, taxing, policing and eminent domain powers, an excellent bond rating, and a dedicated and effective political leadership and staff (including over 300 degreed engineers and scientists of various disciplines). It operates and maintains the facilities it designs and builds, and it has the technical, economic, and political cooperation of local, regional, State and federal agencies and the general support of the public it serves. It has also established a reputation for conceiving and carrying out highly innovative and effective civil works programs and projects.

The History of the Darcy-Weisbach Equation for Pipe Flow Resistance

Glenn O. Brown[1]

Abstract

The historical development of the Darcy-Weisbach equation for pipe flow resistance is examined. A concise examination of the evolution of the equation itself and the Darcy friction factor is presented from their inception to the present day. The contributions of Chézy, Weisbach, Darcy, Poiseuille, Hagen, Prandtl, Blasius, von Kármán, Nikuradse, Colebrook, White, Rouse and Moody are described.

Introduction

What we now call the Darcy-Weisbach equation combined with the supplementary Moody Diagram (Figure 1) is the accepted method to calculate energy losses resulting from fluid motion in pipes and other closed conduits. When used together with the continuity, energy and minor loss equations, piping systems may be analyzed and designed for any fluid under most conditions of engineering interest. Put into more common terms, the Darcy-Weisbach equation will tell us the capacity of an oil pipeline, what diameter water main to install, or the pressure drop that occurs in an air duct. In a word, it is an indispensable formula if we wish to engineer systems that move liquids or gasses from one point to another.

The Darcy-Weisbach equation has a long history of development, which started in the 18th century and continues to this day. While it is named after two great engineers of the 19th century, many others have also aided in the effort. This paper will attempt the somewhat thorny task of reviewing the development of the equation and recognizing the engineers and scientists who have contributed the most to the perfection of the relationship. Many of the names and dates are well known, but some have slipped from common recognition. As in any historical work, others may well find this survey lacking in completeness. However, space limitations prevent an exhaustive commentary, and the author begs tolerance for any omissions. As a final note, to minimize confusion, standardized equation forms and variable symbols are used instead of each researcher's specific nomenclature. Likewise, simple replacements, such as diameter for radius, are made without note.

---

[1] Professor, Biosystems and Agricultural Engineering, Oklahoma State University, Stillwater, OK 74078; phone 405-744-8425; gbrown@okstate.edu

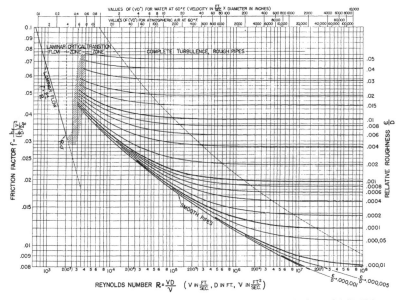

Figure 1. Moody diagram. (Moody, 1944; reproduced by permission of ASME.)

The Equation

The fluid friction between two points in a straight pipe or duct may be quantified by the empirical extension of the Bernoulli principle, properly called the energy equation,

$$h_l = \left(\frac{V_1^2}{2g} + \frac{p_1}{\rho g} + z_1\right) - \left(\frac{V_2^2}{2g} + \frac{p_2}{\rho g} + z_2\right) \approx \left(\frac{p_1}{\rho g} + z_1\right) - \left(\frac{p_2}{\rho g} + z_2\right) \tag{1}$$

where h_l is the fluid friction or head loss between positions subscripted 1 and 2, V is the average velocity, g is the acceleration of gravity, p is the fluid pressure, ρ is the fluid density and z is the elevation of the pipe. When analysis is limited to uniform (constant area) flow, the velocity terms cancel, and the RHS is used. Note that Eq. 1 is not predictive unless all variables on the RHS are known. We must measure pressures in a given pipe system at a specific flow rate to compute the losses. That is, we have to build the system to determine how it will work.

Engineering design requires a relationship that predicts h_l as a function of the fluid, the velocity, the pipe diameter and the type of pipe material. Julius Weisbach (1806-1871) a native of Saxony, proposed in 1845 the equation we now use,

$$h_l = \frac{fL}{D}\frac{V^2}{2g} \tag{2}$$

where L is the pipe length, D is the pipe diameter, and f is a friction factor (Weisbach, 1845). Note that Eq. 2 only predicts the losses due to fluid friction on the pipe wall and the effect of fluid viscosity and does not include minor losses at inlets, elbows and other fittings. While Weisbach presented Eq. 2 in the form we use today, his relationship for the friction factor was expressed as,

$$f = \alpha + \frac{\beta}{\sqrt{V}}$$ (3)

where α and β are friction coefficients that could vary by pipe diameter and wall material. Equation 3 was based on a relatively small data set. Weisbach reported 11 of his own experiments, while 51 measurements were taken from the reports of Claude Couplet (1642-1722), Charles Bossut (1730-1799), Pierre Du Buat (1734-1809), Gaspard Riche de Prony (1755-1839) and Johann Eytelwein (1764-1848).

Weisbach's publication covered most of engineering mechanics and arguably set the standard for all later engineering textbooks. By 1848 it was translated and published in America; a rather remarkable pace for the time. However, his text had no apparent impact in France, the contemporary center for hydraulic research. This is a curious situation since it is believed that Weisbach's interest in hydraulics developed after visiting the Paris Industrial Exposition in 1839 and that he also attended the 1855 Paris World Exposition. Perhaps since Weisbach's equation was based mostly on their data, the French may have believed it provided no improvement over the Prony equation in wide use at the time,

$$h_l = \frac{L}{D}\left(aV + bV^2\right)$$ (4)

where a and b are empirical coefficients. While the exact values of the Prony coefficients were debated, it was believed that they were not a function of the pipe roughness.

A noteworthy difference between Eqs. 2 and 4 is that Weisbach developed a dimensionally homogenous equation. Consequently, f is a non-dimensional number and any consistent unit system, such as SI or English Engineering may be used. That is not the case with Prony's. The roughness coefficients, a and b take on different values depending on the unit system. To the modern eye, Prony's dimensionally inhomogeneous equation is unsightly, but in 1840 there were no electronic calculators, and in fact the modern slide rule was yet to be developed. Thus, Prony's relationship that requires six math operations had a practical advantage over Weisbach's that required eight. Additionally, it was standard practice for the French to drop the first order velocity term, (aV) of Prony's equation to yield a roughly equivalent relationship to Weisbach's that required only four math operations (D'Aubuisson, 1834).

While Weisbach was ahead of most other engineers, his equation was not without precedent. About 1770, Antoine Chézy (1718-1798) published a proportionally based on fundamental concepts for uniform flow in open channels,

$$V^2 P \propto A S$$ (5)

where P is the wetted perimeter, S is the channel slope, and A is the area of flow. Chézy allowed that the proportionality might vary between streams. It is a simple matter to insert a proportionality coefficient, C to yield,

$$V = C\sqrt{RS}$$ (6)

where R is the hydraulic radius given by, $R = A/P$. By introducing the geometry of a circular pipe and noting that under uniform flow conditions $S = h_l/L$, Eq. 6 is transformed to,

$$h_l = \frac{4}{C^2} \frac{L}{D} V^2 \qquad (7)$$

Equation 7 may be considered a dimensionally inhomogeneous form of Eq. 2. Equating one to the other shows that $\dfrac{1}{\sqrt{f}} = \dfrac{C}{\sqrt{8g}}$. Unfortunately, Chézy's work was lost until 1800 when his former student, Prony published an account describing it. Surprisingly, the French did not continue its development, but it is believed that Weisbach was aware of Chézy's work from Prony's publication (Rouse and Ince, 1957).

The Darcy-Weisbach equation (Eq. 2) is considered a rational formula since basic force balance and dimensional analysis dictate that $h_l \propto L\ D^{-1}\ V^2\ g^{-1}$. However, the friction factor, f is a complex function of the pipe roughness, pipe diameter, fluid kinematic viscosity, and velocity of flow. That complexity in f, which results from boundary layer mechanics, obscures the valid relationship and led to the development of several irrational, dimensionally inhomogeneous, empirical formulas. Weisbach deduced the influence of roughness, diameter and velocity on f, but the professional community apparently ignored his conclusions. In addition, the effect of fluid properties was habitually neglected since water at normal temperatures was the only major concern. It would take almost a hundred years and the input of several others for f to be defined completely.

Laminar Flow

By the 1830's the difference between low and high velocity flows was becoming apparent. Independently and nearly simultaneously, Jean Poiseuille (1799-1869) and Gotthilf Hagen (1797-1884) defined low velocity flow in small tubes (Hagen, 1839; Poiseuille, 1841). In modern terms they found,

$$h_l = 64\nu \frac{L}{D^2} \frac{V}{2g}, \qquad (8)$$

where ν is the fluid kinematic viscosity. Note however that neither Poiseuille nor Hagen used an explicit variable for the viscosity, but instead developed algebraic functions with the first and second powers of temperature. The most important aspect of Poiseuille's and Hagen's results was their accuracy. While the restriction to small tubes and low velocity was realized, theirs were the first fluid-friction equations to achieve modern precision. When compared to one another, Hagen's work was more theoretically sophisticated, while Poiseuille had the more precise measurements and looked at fluids other than water. An analytical derivation of laminar flow based on Newton's viscosity law was not accomplished until 1860 [Rouse and Ince, 1957].

Darcy (1857) also noted the similarity of his low velocity pipe experiments with Poiseuille's work. "Before seeking the law for pipes that relates the gradient to the velocity, we will make an observation: it appears that at very-low velocity, in pipes of small diameter that the velocity increases proportionally to the gradient."

Later he showed explicitly that his newly proposed pipe friction formula would reduce to Poiseuille's at low flow and small diameters. He noted that this was a "... rather remarkable result, since we arrived, Mr. Poiseuille and I, with this expression, by means of experiments made under completely different circumstances."

Osborne Reynolds (1842-1912) described the transition from laminar to turbulent flow and showed that it could be characterized by the parameter,

$$\mathbf{Re} = \frac{VD}{\nu} \tag{9}$$

where **Re** is now referred to as the Reynolds number (Reynolds, 1883). The most widely accepted nominal range for laminar flow in pipes is **Re** < 2000, while turbulent flow generally occurs for **Re** > 4000. An ill-defined, ill-behaved region between those two limits is called the critical zone. Once the mechanics and range on laminar flow was well established, it was a simple matter to equate Eqs. 4 and 9 to provide an expression for the Darcy f in the laminar range,

$$f = \frac{64}{\mathbf{Re}} \tag{10}$$

It is unknown who was the first person to explicitly state Eq. 10, but it appeared to be commonly recognized by the early 1900's. Equation 10 is plotted on the left side of Figure 1.

Turbulent Flow

In 1857 Henry Darcy (1803-1858) published a new form of the Prony equation based on experiments with various types of pipes from 0.012 to 0.50 m diameter over a large velocity range (Darcy, 1857). His equation for new pipes was,

$$h_l = \frac{L}{D}\left[\left(\alpha + \frac{\beta}{D^2}\right)V + \left(\alpha' + \frac{\beta'}{D}\right)V^2\right] \tag{11}$$

where α, β, α' and β' are friction coefficients. Darcy noted that the first term on the RHS could be dropped for old rough pipes to yield,

$$h_l = \frac{L}{D}\left(\alpha'' + \frac{\beta''}{D}\right)V^2 \tag{12}$$

where the coefficients α'' and β'' would have different values than for new pipes. Contrary to existing theory, he showed conclusively that the pipe friction factor was a function of both the pipe roughness and pipe diameter. Therefore, it is traditional to call f, the "Darcy f factor", even though Darcy never proposed it in that form.

J. T. Fanning (1837-1911) was apparently the first to effectively combine Weisbach's equation with Darcy's better estimates of the friction factor (Fanning, 1877). Instead of attempting a new algebraic expression for f, he simply published tables of f values taken from French, American, English and German publications, with Darcy being the largest source. A designer could then simply look up an f value from the tables as a function of pipe material, diameter and velocity. However, it should be noted that Fanning used the hydraulic radius, R instead of D in the friction equation. Thus "Fanning f" values are only ¼ of "Darcy f" values. The Fanning form of the equation remains in use in some fields, such as heat exchanger design, where non-circular conduits are common.

During the early 20th century, Ludwig Prandtl (1875-1953) and his students Theodor von Kármán (1881-1963), Paul Richard Heinrich Blasius (1883-1970) and Johann Nikuradse (1894-1979) attempted to provide an analytical prediction of the friction factor using Prandtl's new boundary layer theory. Apparently, Blasius (1913) was the first person to apply similarity theory to establish that f is a function of the Reynolds number. From experimental data he found for smooth pipes,

$$f = \frac{0.3164}{Re^{1/4}} \tag{13}$$

which is now referred to as the Blasius formula and is valid for the range 4000 < **Re** < 80,000. Using data from Nikuradse, the entire turbulent flow range is better fit by the relationship,

$$\frac{1}{\sqrt{f}} = 2 \log\left(Re\sqrt{f}\right) - 0.08 \tag{14}$$

Equation 14 has been referred to both as von Kármán's (Rouse, 1943) and Prandtl's (Schlichting, 1968). It is plotted on Figure 1 and labeled "Smooth Pipes".

Rough pipes offered additional challenges. At high Reynolds number in rough pipes, f becomes a constant that is only a function of the relative roughness, ε/D, where ε is the height of the interior pipe roughness. Similar to the smooth pipe formula, von Kármán (1930) developed an equation confirmed by data collected by Nikuradse (1933),

$$\frac{1}{\sqrt{f}} = 1.14 - 2 \log\left(\frac{\varepsilon}{D}\right) \tag{15}$$

The horizontal lines on the right of Figure 1 plot Eq. 15 for various ratios of ε/D.

The transition region between laminar and fully turbulent rough pipe flow was defined empirically by detailed measurements carried out by Nikuradse (1933) on pipes that had a uniform roughness created by a coating of uniform sand covering the entire pipe interior. His data showed clear trends that could be explained by the interaction of the pipe roughness with the fluid boundary layer. However, measurements by Colebrook and White (1937) showed that pipes with non-uniform roughness did not display the same transition curves. For commercial pipes White (1939) showed the transition region could be described by,

$$\frac{1}{\sqrt{f}} = 1.14 - 2 \log\left(\frac{\varepsilon}{D} + \frac{9.35}{Re\sqrt{f}}\right) \tag{16}$$

Equation 16 is plotted in Figure 1 for various ratios of ε/D in the region labeled "Transition Zone".

Integration

It would wait for Hunter Rouse (1906-1996) in 1942 to integrate these various formulas into a useful structure. He noted unambiguously, (Rouse, 1943) "These equations are obviously too complex to be of practical use. On the other hand, if the function which they embody is even approximately valid for commercial surfaces in general, such extremely important information could be made readily available in diagrams or tables." Using published data he showed Eq. 16 was a reasonable function for commercial pipe. Rouse then developed a diagram (Figure 2) that presented Eqs. 10, 14, 15, and 16 in a systematic and somewhat intricate fashion. The primary vertical axis plotted $1/\sqrt{f}$, the primary horizontal axis plotted $\mathbf{Re}\sqrt{f}$, and secondary axes plotted $\mathbf{Re}$ and f. Equations 15 and 16 were plotted for various values of relative roughness. In an open corner, he also provided a convenient list of pipe roughness. Rouse's original contribution in addition to the overall synthesis was defining the boundary between the transition and fully turbulent zones as,

$$\frac{1}{\sqrt{f}} = \frac{\varepsilon}{D} \frac{\mathbf{Re}}{200} \qquad (17)$$

Equation 17 is plotted on both Figures 1 and 2.

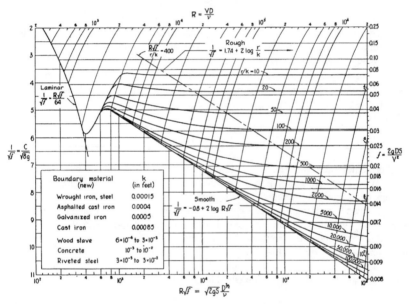

Figure 2. Rouse diagram. (Rouse, 1943; reproduced by permission of IIHR.)

Lewis Moody (1880-1953) was in the audience when Rouse presented his paper. Moody felt that Rouse's diagram was "inconvenient" and decided to redraw Rouse's diagram "in the more conventional form used by Pigott, ..." (Moody, 1944). Moody's paper was primarily an instructional lecture, and as he said, "The author does not claim to offer anything particularly new or original, his aim merely being to embody the now accepted conclusion in convenient form for engineering use." Moody acknowledged previous researchers, and reproduced figures from both Colebrook and Rouse.

It should be noted that Moody's diagram is more convenient to use when finding h_l with known Q and D. However, Rouse's diagram allows a direct, non-iterative solution for Q with known h_l and D. Thus, each has its advantages.

A Rose by Any Other Name

The naming convention of the Darcy-Weisbach's equation in different countries and through time is somewhat curious and may be tracked in the contemporaneous textbooks. Generally, French authors have identified any relationship in the form of Eqs. 2 or 4 as "La formule de Darcy". The friction factor may be listed as either f or as Darcy's Number, **Da**. An early English text, (*Neville*, 1853) identified Eq. 2 as the *"Weisbach Equation"*, but later authors have generally adopted the French terminology. Surprisingly up to the 1960's, German authors either gave it a generic name like "Rohrreibungsformel" (pipe formula) or use the French jargon. However, almost all German authors now use "Darcy-Weisbach".

The equation's designation has evolved the most in America. Early texts such as Fanning (1877) generally do not name the equation explicitly; it is just presented. In the period around 1900 many authors referred to Eq. 2 as Chézy's or a form of Chézy's (Hughes, and Stafford, 1911). However, by the mid-century, most authors had again returned to leaving the equation unspecified or gave it a generic name. Rouse in 1942 appears to be the first to call it "Darcy-Weisbach" (Rouse, 1943). That designation gained an official status in 1962 (ASCE, 1962), but did not become well accepted by American authors until the late 1980's. A check of ten American fluid-mechanics textbooks published within the last eight years showed that eight use the Darcy-Weisbach naming convention, while two continue to leave the equation's name unspecified. While variations across oceans and languages are to be expected, it is disappointing that a single nomenclature for Eq. 2 has not been adopted after 157 years, at least in the United States.

Rather ironically and contrarily to the practice with the equation name, the f versus **Re** diagram is universally credited to Moody, and the contributions of others are seldom acknowledged. This was a sore point with Rouse (1976), and he wrote of their 1942 meeting (in third person),

"After the Conference, Lewis Moody of Princeton suggested using the latter variables (f and **Re**) as primary rather than supplementary, as in the past, but Rouse resisted the temptation because he felt that to do would be a step backward. So Moody himself published such a plot, and it is known around the world as the Moody diagram!"

In his writing, Rouse used the exclamation point very sparingly, thus the intensity of his opinion is apparent.

Closing Comments

With Moody's publication, practitioners accepted the Darcy-Weisbach equation and it is dominant in most engineering fields. Its results are applied without question, which may not be appropriate for all conditions. Rouse (1943) showed that the Eq. 16 is only valid for pipes with interior roughness similar to iron. Spiral or plate fabricated pipes had substantially different functions. Statements of true accuracy are rare, but based on his personal judgment that pipe roughness is difficult to define, White (1994) has stated the Moody chart is only accurate to $\pm 15\%$. Thus, it is surprising that the diagram has not been modified or replaced over the last 58 years. Efforts have been made to streamline the procedure and eliminate the manual use of graphs. This difficultly is responsible for the continued use of less accurate empirical formulas such as the Hazen-Williams equation. While they have a limited Reynolds number range, those irrational formulas are adequate for some design problems. Therefore, the most notable advance in the application of the Darcy-Weisbach equation has been the publication by Swamee and Jain (1976) of explicit equations for pipe diameter, head loss and the discharge through a pipe, based on the Colebrook-White equation. Swanee and Jain's formulas eliminate the last advantages of the empirical pipe flow equations. Thus, because of its general accuracy and complete range of application, the Darcy-Weisbach equation should be considered the standard and the others should be left for the historians. Liou (1998) presented an interesting discussion on the topic.

By necessity this was a brief survey of the historical development that focused solely on the Darcy-Weisbach equation and the Darcy friction factor, f. Additional theoretical background on f is presented by Schlichting (1968), while an excellent historical overview that includes other pipe friction formulas is provided by Hager (1994).

References

ASCE (1962). *Nomenclature for hydraulics*. ASCE, New York, 501 pages.

d'Aubuisson, J. F. (1834). *Traité de l'hydraulique à l'usage des ingénieurs*, Pitois-Levrault, Paris (in French).

Blasius, H. (1913). *Das Ähnlichkeitsgesetz bei Reibungsvorgängen in Flüssigkeiten*, Forschungs-Arbeit des Ingenieur-Wesens 131. (in German).

Colebrook, C. F. (1939) "Turbulent flow in pipes with particular reference to the transition region between the smooth and rough pipe laws." *Proc. Institution Civil Engrs.*, 12, 393-422.

Colebrook, C. F. and White, C. M. (1937). "Experiments with fluid-friction in roughened pipes." *Proc. Royal Soc. London*, 161, 367-381.

Darcy, H. (1857). *Recherches expérimentales relatives au mouvement de l'eau dans les tuyaux*, Mallet-Bachelier, Paris. 268 pages and atlas (in French).

Fanning, J. T. (1877). *A practical treatise on water-supply engineering*, Van Nostrand, New York, 619 pages.

Hagen, G. (1839). "Über die Bewegung des Wassers in engen zylindrischen Röhren." *Pogg. Ann.*, 46, 423-442 (in German).

Hager, W. H. (1994). "Die historische Entwicklung der Fliessformel." *Schweizer Ingenieur und Architekt*, 112(9), 123-133. (in German)

Hughes, H. J. and Stafford, A. T. (1911). *A treatise on hydraulics*, Macmillan, New York. 501 pg and 5 plates.

Kármán, Th. von (1930). "Mechanische Aehnlichkeit und Turbulenz." *Proc. Third International Congress for Applied Mechanics*, C. W. Oseen and W. Weibull eds., Stockholm. Vol. 1, 79-93 (in German).

Liou, C. P. (1998). "Limitations and proper use of the Hazen-Williams equation." *J. Hydraulic Engrg.*, ASCE 124(9), 951-954.

Moody, L. F. (1944). "Friction factors for pipe flow." *Trans. ASME*, 66:671-678.

Neville, John, (1853). *Hydraulic tables, coefficients and formulae*. John Wale, London, 224 pages.

Nikuradse, J. (1933). "Strömungsgesetze in rauhen Rohren." *Forschungs-Arbeit des Ingenieur-Wesens* 361 (in German).

Poiseuille, J. L. (1841). "Recherches expérimentales sur le mouvement des liquides dans les tubes de très-petits diamètres." *Comptes Rendus*, Académie des Sciences, Paris 12, 112 (in French).

Reynolds, O. (1883). "An experimental investigation of the circumstances which determine whether the motion of water shall be direct or sinuous and of the law of resistance in parallel channel." *Phil. Trans. of the Royal Soc.*, 174:935-982.

Rouse, H. (1943). "Evaluation of boundary roughness." *Proceedings Second Hydraulics Conference*, Univ. of Iowa Studies in Engrg., Bulletin No. 27.

Rouse, H. (1976). *Hydraulics in the United States, 1776-1976*, Iowa Institute of Hydraulic Research, Univ. of Iowa, Iowa City, 238 pg.

Rouse, H. and Ince, S. (1957). *History of Hydraulics*, Iowa Institute of Hydraulic Research, Univ. of Iowa, Iowa City, 269 pg.

Schlichting, H. (1968). *Boundary-layer theory*, Translated by J. Kestin, 6th ed, McGraw-Hill, New York, 748 pg.

Swamee, P. K., and Jain, A. K. (1976). "Explicit equations for pipe-flow problems." *J. Hydraulics Division*, ASCE, 102(5), 657-664.

Weisbach, J. (1845). *Lehrbuch der Ingenieur- und Maschinen-Mechanik, Vol. 1. Theoretische Mechanik*, Vieweg und Sohn, Braunschweig. 535 pages (in German).

White, F. M. (1994). *Fluid mechanics*, 3rd ed., Mc-Graw Hill, New York. 736 pg.

Manning's Formula by Any Other Name

Jerry L. Anderson[1]

Abstract

The history of the development of the open channel flow resistance equations is reviewed. A detailed discussion of Philippe-Gaspard Gauckler's contribution to the development of an open channel flow resistance equation is contrasted with the subsequent development of a similar equation by Robert Manning. Both men were practicing professional engineers and were desirous of a means of easily calculating the average stream velocity as a function of channel geometry, slope, and channel boundary characteristics. As practicing engineers, both were still conscious of the basic fundamental principle of fluid mechanics that today is called uniform steady flow. Both used the same data from Darcy and Bazin and arrived at a similar equation without the knowledge of each other's separate development. Consequently, recognition should be afforded to Philippe-Gaspard Gauckler by referring to the well known and popular open channel flow resistance equation known today as Manning's equation as the Gauckler-Manning's equation.

Introduction

Rouse and Ince (1963) in *History of Hydraulics* credit Aristotle (384-322 B.C.) with observing two types of action with regard to the movement of one element through another; a motivation, to which the speed of movement is directly proportional, and a resistance, to which it is inversely proportional. So, suffice it to say that the interest in open channel flow resistance has perplexed philosophers, mathematicians, hydraulicians, and engineers since Aristotle, but to the credit of our forefathers the challenge has been meet. The credit for offering the first open channel flow resistance equation and the most lasting formula is given the French engineer, Antoine Chezy (Rouse and Ince, 1963). Since Chezy first introduced his simple formulation of velocity as a function of slope, area, and perimeter of the section of the

----

[1]F.ASCE, PhD, Director, Ground Water Institute, The University of Memphis, Memphis, TN 38152-3170, jlandrsn@memphis.edu, phone (901) 678-3062, fax (901) 678-3078

current touching the boundaries of the channel, various mathematicians, hydraulicians, and engineers have offered different equations, approaches, and methods for incorporating the loss of energy, either kinetic or potential, in the open channel flow phenomena. Engineers have proposed different and various means of incorporating the mean velocity of flow as a function of hydraulic radius and energy gradient slope. Williams (1970) reveals that at least ten such proposals were forthcoming after of Darcy and Bazin's (1865) and Humphrey and Abbot's (1861) data became available. The most widely used equation in English-speaking countries today that relates velocity to channel properties, hydraulic radius, and slope of the energy gradient is called Manning's equation as denoted in (1) below.

$$V = \frac{k_n}{n} R_h^{\frac{2}{3}} S^{\frac{1}{2}}$$ (1)

where

V = mean velocity
k_n = 1.486 for English units, and 1 for SI units
n = Manning's roughness coefficient
R_h = hydraulic radius
S = slope of the energy grade line, h_l/L
h_l = energy loss in the reach distance of L
L = river reach distance between two section of interest

Fischenich (2000) notes that the original equation was presented with k_n/n in (1) represented as simply C. Even Manning was not confident of this equation as Powell (1968) notes, where the equation for the resistance to the flow of water in open channels that Manning first recommended in his address to the Institution of Civil Engineers 1889 is entirely different from the equation that bears his name today. Manning notes that his independent formulation of (1) had previously been deduced by G.H.L. Hagen (1881). Further, Powell notes that Gauckler (1867, 1868) predates Hagen with his equations. Williams (1970) cites at least ten different equations relating the average channel velocity to a function of the hydraulic radius and slope. Williams does not distinguish between the slope of the energy grade line and the slope of the channel bottom. However, the popularity of calling the open channel flow resistance equation Manning's equation is probably due to Flamant (1891). Williams (1970) states that the first person to start using the expression Manning's formula in regard to equation (1) was A. A. Flamant, Engineer-in-chief of the Ponts et Chaussées, in his 1891 book. Powell contends that Bovey (1901) was the first to put the equation into its English form, as we know it today. So it is appropriate that we review the historical development of the resistance equation for open-channels.

The two primary contributors to a form of the equation that is today called Manning's equation are Robert Manning (1816-1897) and Philippe Gauckler (1826-1905). Sometimes Albert Strickler's (1887- 1963) name is associated with the flow resistance equation, but his contributions were made in 1923 at a much later date. Actually, W. H. Hager (2001) does give Strickler some credit and refers to the

Gauckler-Manning-Strickler (GMS) formula as currently the most popular power formula for estimating the relationship between average cross-sectional velocity, hydraulic gradient, and hydraulic radius, for both closed-conduit and uniform open channel flows. However, in this paper only Gauckler's and Manning's contributions will be reviewed.

Philippe-Gaspard Gauckler

A debt is due to Willi H. Hager (2001) for his interest in the contribution by Philippe Gauckler. Had it not been for his dedication and persistence for discovering more about Gauckler's life and career, what we know today of Gauckler's contribution to the open channel flow resistance equation may have been lost forever. Hager contributes the advancement on the understanding of open channel flow resistance to two developments that took place in the 1860s. The first work was attributed to data collected by two Topographic Engineers and reported in 1861. Humphreys and Abbot (1861) were charged with conducting a comprehensive topographical and hydrographical survey and reporting upon the physics and hydraulics of the Mississippi River, upon the protection of the alluvial region against overflow, and upon the deepening of the mouths. Humphreys states that while the primary purpose of this work was intended to render a solution to the problem of the protection of the alluvial region of the Mississippi against inundation, it was intended as well to determine the laws governing the flow of water in natural channels and to develop means of expressing them in formulas that could safely and readily be used in practical applications. Secondly, Henry Philibert Gaspard Darcy (1803-1858) and Henri Emile Bazin (1829-1917) presented data from their detailed observations (Darcy and Bazin, 1865) collected from a variety of flow configurations in the Canal de Bourgogne. Since the data sets from these two publications were varied for different flow and channel conditions, further analysis could be done (Hager, 2001).

As Hager reports, Gauckler published his main paper in 1868 after noting the formulas proposed by Dubuat, Prony, and Eytelwein. Rouse and Ince (1963) reported the recommendation of two equations by Gauckler, which were dependent on the slope of the channels. Actually, the equation was of one form as shown below, but could be simplified based on channel slopes.

$$V^{\frac{1}{2}} + R_h V^{\frac{1}{4}} = \alpha (4R_h)^{\frac{1}{3}} S^{\frac{1}{4}} \tag{2}$$

As reported by Hager, this formula comprises the sum of two power expressions in velocity; α is a coefficient describing boundary roughness and typically varied between 5 and 7. However, Gauckler noted for small slopes (S<0.0007), the first term on the left side of (2) could be dropped to give (3) below

$$V^{\frac{1}{4}} = \beta R_h^{\frac{1}{3}} S^{\frac{1}{4}} \tag{3}$$

For slopes $S \geq 0.0007$, the second term in (2) may be dropped to give the following expression.

$$V^{\frac{1}{2}} = \alpha R_h^{\frac{1}{3}} S^{\frac{1}{4}}$$
(4)

In (3) and (4) above, both α and β are roughness coefficients. It is important to note at this time as correctly reported by Powell (1968) that the "less than" and "greater than" symbols in Rouse and Ince (1963) were interchanged in their equations on page 179. Williams (1970) reported that Gauckler proposed (4), on the basis of Darcy and Bazin (1865) data. The results of Darcy and Bazin's experiment have been considered to be one of the great works of the 19th century engineering research. These open channel flow experiments were conducted in artificially constructed canals lined with cement, wood, brick, gravel and rock and were of rectangular, trapezoidal, triangular and semicircular cross sections. Also (4) may be expressed in the current form of (1) if

$$\alpha = \frac{k_n}{n}$$
(5)

Hager notes that there are four reasons why (4) was not widely adopted. The general flow equation (2) was subdivided arbitrarily into two regions, depending on slope, rather than flow regimes. He notes that the flow regimes of laminar and turbulent flow were not clearly understood at that time. Engineers questioned what was the mechanism responsible for the abrupt modification from (4) to (5). Secondly, the simplicity of (1) was questioned by engineers of that day whereas the more cumbersome equation suggested by Emile Ganguillet (1818-1894) and Wilhelm (1818-1888) looked more reliable. Hager notices that thirdly there was a small awkwardness in that, in two papers, Gauckler's name was inaccurately written as "Gauchler". And, fourthly, Gauckler quit working in the field of hydraulics after his contribution in the 1860s.

Gauckler was what we call today a practicing engineer in that most of his work was done for municipalities and the French government. Gauckler was schooled in Strasbourg and graduated from the Ecole Polytechnique in Paris. He graduated in 1848 and began working for the Ponts et Chaussées Corps. He undertook various engineering works, but it was not until the earlier 1860s that he took on a Ponts et Chaussées appointment that required regulating flow in the Rhine River and directing the imperial fish-hatching institution of Huningue. He improved the water supply system for the city of Colmar. Hager describes many projects that Gauckler was involved with and how the projects sparked his interest in understanding open channel flow resistance. It was after he was promoted to engineer first class in the Ponts et Chaussées that he published his papers on water flow and open channel resistance.

The year 1870 marked a great change for France and Gauckler. France lost Alsace to Germany as a result of the German-French War and Gauckler was involved in war fortifications at Mont Rolland and involved in the water supply of the beleaguered city of Belfort. Subsequently, in 1881 he was promoted to director of state railways and received a promotion to chief engineer first class. Gauckler retired in 1891, but had spent only a short time in the 1860s involved in water resources projects. Hager suspects that Gauckler is not well recognized for his contribution to the resistance of flow in open channels for two reasons: primarily due to his short professional period in hydraulics; and secondly, he was not associated with a university.

Robert Manning

Credit is given to Chow (1955), Powell (1968), Williams (1970), Dooge (1989), and Fischenich (2001) for uncovering and reporting much of the history of the Manning formula and Robert Manning's contribution to its development. Williams (1970) reports that Robert Manning published a form of the equation that was named after him and avowed that it was applicable to everything from pipes ½ inch in diameter to the Mississippi River.

Robert Manning was born in Normandy the year following the battle of Waterloo (1816) and moved with his mother to Waterford, Ireland, in 1826 following the death of his father (Fischenich, 2000). Although he briefly worked as an accountant for his uncle, John Stephans, he was drafted into the Arterial Drainage Division of the Irish Office of Public Works in 1846. In 1874, Manning was appointed as chief engineer of the Office of Public Works where he was responsible for various arterial drainage, inland navigation, sewerage, and harbor projects. Also, he served a term as president of the Institution of Civil Engineers of Ireland (Rouse and Ince, 1963). Manning wrote his first paper on hydraulics in 1851 and his last paper in 1895 (Dooge, 1989).

Manning (Dooge, 1989 and Fischenich, 2000) began his search for a formula relating the average velocity of flow in an open channel to the geometry of the channel, slope of the channel, and an indicator of the channel boundary influence by evaluating and comparing the seven best-known open channel formulae of the time. The seven formulae chosen were those by Du Buat, Eytelwein, Weisbach, St. Venant, Neville, Darcy and Bazin, and Ganguillet and Kutter. Manning's understanding of the balance of forces in uniform flow is cited in his first paper (Manning, 1851) as follows:

> "First – In uniform motion the accelerating and retarding forces balance each other; and, secondly, that the resistance which retards the motion of water in a river channel is proportion to the length of the wet peremiter (sic) to the square of the velocity, and is the inverse ratio of the section."

This is the classical statement of the principal of uniform flow as reported by Chezy and Du Buat. Manning in his position as engineer for the Arterial Drainage Works must have been seeking a practical equation to relate flow rate to the measurable channel properties. Manning (1852) used (6) below to develop tables in order to determine the discharge and dimension of river courses. Dooge (1989) noted that

Manning was a busy professional engineer with a respect for hydraulic theory not only for its own sake but also as the means for avoiding mistakes in practical engineering decisions.

$$Q = 55A\sqrt{2DR}_h \tag{6}$$

where Q = flow rate, ft^3/min
   D = declivity or fall in feet per mile, ft/mile
   A = area of transverse section, ft^2,
   R_h = mean hydraulic depth, or the area of the transverse section divided by the perimeter in contact with the fluid. Dooge (1989) cites Manning's definition of R_h as the mean hydraulic depth, but this clearly is the definition of the hydraulic radius that is commonly used today.

Manning was always conscious of the need to understand the development of equations and the limitations for which they were developed. Both Dooge and Fischenich cite a portion of Manning's Presidential Address to the Institution of Civil Engineers of Ireland (Manning, 1878):

"And now a few words, again addressed to the younger members, with regard to the use of formulae. I very much fear that if I were to illustrate any observations I have to make with chalk on the black-board, a dozen note-books would be taken out, the formula copied without investigation, probably to my discredit, and eventually (sic) worked to death. It should be remembered that a formula is only a short memorandum (put in a shape fit for ready use) of the result arrived at after a patient consideration of the facts and principles upon which it is founded and to use it without investigation is the merest empiricism. I have used this last word in its popular, and not in its mathematical sense. There are no formulae so useful – I might almost say none of real value – to the engineers, which are not in some degree empirical, embodying facts arrived at by careful experiment."

Dooge (1989) cites the first formula found by Manning to fit the mean V-R_h relationship was:

$$V = 32\sqrt{R_h S(1 + R_h^{\frac{1}{3}})} \tag{7}$$

which Manning described as "entirely empirical". Further, Manning suggested the exponent of S was a constant and equal to the square of that function. Subsequently, it was possible to represent the velocity by a monomial equation similar to Chezy's, and it took the form shown below:

$$V = CR_h^{\ x} S^{\frac{1}{2}} \tag{8}$$

Using (8) as the basis for determing the average velocity in an open channel, Manning then availed himself of the data of Bazin's experiments on artificial channels specially constructed for experimental purposes. Using the results Manning obtained from this data, he considered it to be sufficiently accurate to take the value of "x" in (8) to be 0.666 or 2/3. Hager (1988) recognized that the exponent could vary between 0.60 and 0.75 depending on the relative boundary roughness, but stated for typical civil-engineering values of relative roughness $10^{-3} < \varepsilon < 10^{-1}$, where $\varepsilon = k_s/4R_h$ with k_s = Nikuradse sand roughness, a power close to 2/3 fit best. Apparently without knowledge of Gauckler's proposal, Manning (1889) recommended (9) below, which is equal to (1) if C is identified properly:

$$V = CR_h^{\frac{2}{3}} S^{\frac{1}{2}} \tag{9}$$

Eventually Manning rejected (9) for a more dimensionally homogenous equation (10). Manning, being the practical engineer that he was, considered the extraction of a cube root (Rouse and Ince, 1963) to be inconvenient and tedious and the lack of dimensional homogeneity in the monomial formula (9) was very disconcerting to him. He observed that if a formula has been developed strictly from observations, then the use of that formula could not in strictness be applied to a single case outside them (Rouse and Ince, 1963). Manning was so determined to maintain dimensional homogeneity that he rejected (9) in favor of the dimensionally homogeneous relationship, which incorporates barometric pressure:

$$V = C\sqrt{gS}\left[\sqrt{R} + \frac{0.22}{\sqrt{m}}(R - 0.15m)\right] \tag{10}$$

where m is the barometric pressure in meters of mercury. Dooge states that Manning did not consider (10) being an alternative to (9), but rather two ways of calculating velocity to sufficient accuracy over the range of practical interest. Both forms of the Manning formulas were referred to in subsequent literature, but after about 30 years the references relate only to the monomial form of (9).

Williams (1970) reviewed the development of resistance equations from Foss, Crimp and Bruges, Tutton, Yarnell and Woodward, and Albert Strickler. These equations were developed after the introduction of the monomial equation by Gauckler and Manning. Foss was an assistant engineer for Boston Water Works in 1894 when he introduced his equation that had the same form as (9), but made no reference to Manning, Gauckler, Hagen, Vallot, or Thrupp. Vallot and Thrupp had previously suggested similar formulas of the same form for the open channel flow resistance equation. Crimp and Bruges were from London and both were engineers. Crimp was the former District Engineer to the London County Council. Williams states that they used the data from Darcy and Bazin's experiments and suggested an equation for sewers and iron pipes. Their equation was similar to (9) where C = 124. However, Crimp and Bruges did not mention any of the previous work of the six men who had preceded them. Tutton, an engineer for the Public Works in Buffalo, New York, proposed an equation similar to (9) where C = 1.54/n, in which n = Kutter's n. Tutton actually offered a theoretical development of the resistance equation, which no one else had attempted. Tutton seems to have recognized all of the previous contributions to the development of the resistance equation except Manning's. Yarnell and Woodward (1920), respectively Senior Drainage Engineer for the US Department of Agriculture and Professor of Hydraulics at the University of Iowa, ran tests on drain tile and offered (9) where C = 138. Yarnell and Woodward did not mention the similarity between their formula and many equations of the same form proposed earlier. They reasoned that tile drains were so uneven at the joints and had so many joints that previous equations were not applicable. Albert Strickler (1923) reviewed many of the previously proposed formulas and then applied 17 sets of open channel and pipe data taken by the Bern Eidgenössische Amt für Wasserwirtschaft. He reported an equation similar to (9) where C was referred to as a resistance coefficient. Although his name is occasionally attached to the resistance equation (9), he specifically recognized the previous contributions of Gauckler, Hagen, Manning, and Tutton.

As has been described, the hydraulic developments in the 19th Century with regard to a monomial resistance equation for flow were truly remarkable. All of the researchers seemed to narrow in on the generic form of (9) where C was a factor that included the specific resistance for that specific channel. Researchers have offered various explanations for the value of C, but Manning definitely did not originally recommend using 1/n, in which n is Kutter's n, in place of his C (Williams, 1970). Dooge (1989) cites Manning's own admission of the reciprocal of C corresponding closely with that of n as determined by Ganguillet and Kutter with both C and n being constant for the same channel. Flamant, in his 1891 book, seems to be responsible for the first introduction of 1/n for C. Consequently, Flamant played a major role both in naming the equation Manning's formula and for using Kutter's n to represent the value of the resistance parameter. Williams reports that Tutton in 1896 designed his formula specifically to include Kutter's n and was the first to actually write the equation using n rather than C. Subsequent use of n as the resistance parameter found its way into textbooks and other manuscripts and has become the adopted practice. H.T. Bovey (1901) of McGill University was the first author to use (1) in the English form:

$$V = \frac{1.486}{n} R_h^{\frac{2}{3}} S^{\frac{1}{2}} \tag{11}$$

Williams cites Parker, Daugherty (1916), and King (1918) as following the same pattern of (11) in their texts. It suffices to say that the influence of Daugherty has continued until today in that his original text is in its 10th edition with authors Finnemore and Franzini (2002). King's *Handbook of Hydraulics* is in its 7th edition with authors, Brater, King, Lindell and Wei (1996). Fischenich (2000) provides a litany of modern researchers who have contributed to further understanding of the relationship between flow resistance and fluid mechanics. Rouse (1961) provides a critical analysis of open-channel resistance and suggests that there is a valid basis for the analysis of wide channels as well as circular pipes using the semi-logarithmic velocity-distribution and resistance laws.   Rouse pointed out the effects of cross-sectional shape, boundary nonuniformity, and flow unsteadiness, in addition to viscosity and wall roughness. Yen (2002), in his Hunter Rouse Award Lecture in 1999, extended the work reported by Rouse in 1965. Yen continues by discussing the differences between momentum and energy resistances, between point, cross-sectional and reach resistance coefficients, as well as compound/composite channel resistance.

Conclusion

Fischenich (2000) states that despite the fact that these "new" formulae include many improvements from the standpoint of incorporating fluid mechanic concepts, Manning's monomial formula persists as the most frequently used by hydraulic engineers today. Williams (1970) and Hager (2001) each in his own way make an argument that the equation known as Manning's formula in English–speaking countries should really be called the Gauckler-Manning equation. Williams discounts Hagen's claim since his equation was developed strictly for the Ganges Canal and notes that the Manning's equation was well established by the time of Strickler's 1923 paper. Manning was the first person to mention that the reciprocal of his coefficient C was very nearly the same as Kutter's n. Williams contends that the failure of Manning and others to give Gauckler credit for first presenting the general type-equation should not be allowed to obscure Gauckler's important role in the history of this formula. Consequently, whether one calls it Manning's equation or rightly so Gauckler-Manning's equation, it accomplishes the same purpose.

References

Bovey, H. T. (1901). *A Treatise on Hydraulics,* 2nd Ed., John Wiley & Sons, New York

Brater, E. F., H. W. King, J.E. Lindell, C.Y. Wei (1996). *Handbook of Hydraulics,* 7th Ed., McGraw-Hill, New York

Chow, V. T. (1955). "A Note on the Manning Formula", *Transactions,* American Geophysical Union, 36(4), p 688 with discussions by H. Rouse, J.M. Robertson, J.C.I. Dooge, and V. T. Chow, 37(3), 327-330

Darcy, H. and H. Bazin (1865). "Recerches hydrauliques; 1re partie, recherches expérimentales sur l'écoulement de l'eau dans les canaux découverts; 2° partie, recherches expérimentales relatives aux remous et á la propagation des ondes", *Académie des Sciences*, Paris

Daugherty, R.L. (1916). *Hydraulics*, McGraw-Hill, New York

Dooge, J. C. I. (1989). "The Manning Formula in Context", *Channel Flow and Catchment Runoff*, Proceeding of the International Conference for Centennial of Manning's Formula and Kuichling's Rational Formula, edited by Ben Chie Yen, University of Virginia, Charlottesville, VA

Finnemore, E.J. and J. B. Franzini (2002). *Fluid Mechanics with Engineering Applications*, McGraw-Hill, New York

Fischenich, C. (2000). "Robert Manning – A Historical Perspective", ERDC TN-EMRRP-SR-10, US Army Research and Development Center, Environmental Laboratory, Vicksburg, MS

Flamant, A. A. (1891). *Méchanique appliquée –Hydraulique*, Baudry & Cie, Paris

Gauckler, P.G. (1867). "Études théoriques et pratiques sur l'écoulement et le mouvement des eaux", *Comptes Rendus de L'Academie des Sciences*, 64, 818-822, Paris

Gauckler, P.G (1868). "Du mouvement de l'eau dans les conduites", *Annales des Ponts et Chaussées*, 38 (1), 228-281

Hager, W. H. (1988). "Abflussformeln für turbulente Strömungen", *Wasserwirtschaft*, 78(2), 79-84

Hager, W.H. (2001). "Gauckler and the GMS Formula", *J. Hyd. Div.*, ASCE, 127(8), 635-638

Hagen, G.W. (1881). "Neuere Beobachtung über die gleichförmige Bewegung des Wassers", *Zeitscrhift für Bausesen*, 31, 403 – 408

Humphreys, A. A. and H. L. Abbot (1861). "Report upon the Physics and Hydraulics of the Mississippi River upon the Protection of the Alluvial Region Against Overflow", Corps of Topographical Engineers Professional Papers No.4, United States Army, War Department

King, H.W. (1918). *Handbook of Hydraulics*, 1st Ed. McGraw-Hill, New York

Manning, R. (1851). "Observations on subjects connected with Arterial Drainage", *Transaction of Institution of Civil Engineers of Ireland*, Part II, IV: 90-104

Manning, R. (1852). *Tables for Calculating the Discharge, Inclination and Dimensions of River Courses and Drains,* Thom, Dublin

Manning, R. (1880). Presidential Address to the Institution of Civil Engineers of Ireland, 6th February 1878, *Transactions of Institution of Civil Engineers of Ireland*, 12, 68-85

Manning, R. (1889). "On the flow of water in open channels and pipes", *Transactions of Institution of Civil Engineers of Ireland*, 20, 161-166

Powell, R. (1968). "The Origin of Manning's Formula", *J. Hyd. Div.*, ASCE, 94(4), 1179 –1181

Rouse, H. and S. Ince (1963). *History of Hydraulics*, Dover Publications, Inc, New York

Rouse, H. (1965). "Critical Analysis of Open-Channel Resistance", *J. Hyd. Div.*, ASCE, 91(4), 1-25

Strickler, A. (1923). "Beiträge zur Frage der Geschwindigkeitsformel und der Rauhigkeitszahlen für Ströme, Kanäle, und geshhlossene Leitungen, " *Mitteilungen des Amtes für Wasserwirtschaft,* 16, Bern

Williams, G.P. (1970). "Manning Formula – A Misnomer?", *J. Hyd. Div.*, ASCE, 96(1), 193 – 200

Yarnell, D.L. and S. M. Woodward (1920). "The Flow of Water in Drain Tile", *U. S. Department of Agriculture Bulletin,* No. 854

Yen, B. C. (2002). "Open Channel Flow Resistance, *J. Hyd. Div.*, ASCE, 128(1), 20 – 39

History of Water Distribution Network Analysis: Over 100 Years of Progress

Dhandayudhapani Ramalingam[1], Srinivasa Lingireddy[2], and Lindell E. Ormsbee[3]

Abstract

This paper provides a brief history of water distribution system modeling, starting with use of the Hazen-Williams equation and the Equivalent Pipe Method and continuing to the use of sophisticated computer models. Interesting historical perspectives are provided as well as insights into the application of the actual methods.

Introduction

Computation of flows and pressures in network of pipes has been of great value and interest for those involved with design, construction and maintenance of public water distribution systems. In a recent article, Jesperson (2001) provides a brief historical overview of public water systems. According to Jesperson, the idea of public water supply systems can be traced back to as early as 700 BC when "qanats" (slightly sloped hillside tunnels) that brought water to Persia were built. Romans started constructing aqueducts from 312 BC. Recently discovered spring-water collection systems at Machu Pichu in Peru shed light on the elaborate engineering designs of ancient water supply systems. The Machu Pichu water supply systems date back to 1450 AD. The first public water system in the United States dates back to 1652 AD when the City of Boston incorporated its water works formed to provide water for fire-fighting and domestic use. Since then many public water supply systems came into existence. In the earlier periods channels were made from cut stone, brick, or rubble. Pipes were made mostly from drilled stone, wood, clay and lead. In the 18th century, cast iron pipes replaced the wooden pipes. The 19th century witnessed significant improvements in making pipe joints that withstood high pressures. Water supply pipelines made of steel, ductile-iron, asbestos cement and reinforced concrete came into increasing usage during the early 20th century. Increasing complexities associated with distribution systems necessitated precise estimation of flows and pressures in various parts of the distribution system. Solution of the single-pipe flow problem was no longer adequate. The quest for methods that analyze (solve for flows and pressures of the) entire water distribution network gave birth to the topic "water distribution network analysis" or "pipe network analysis."

1 Doctoral Scholar, Department of Civil Engineering, University of Kentucky, Lexington, Kentucky 40506-0281, 859.257.8005. E-mail : drama0@engr.uky.edu
2 Associate Professor, Department of Civil Engineering, University of Kentucky, Lexington, Kentucky 40506-0281, 859.257.5243. E-mail : lreddy@engr.uky.edu
3 Interim Director, Tracy Farmer Center for the Environment, University of Kentucky, Lexington, Kentucky 40506-0107, 859.257.6329. E-mail : lormsbee@engr.uky.edu

Early Perspectives on Network Analysis

As a discussion to one of the notable publications on water distribution network analysis, Stanley (1937) succinctly puts the importance given to network analysis methods in those days by saying *"Methods of design for transmission and distribution systems have had too little attention from water works engineers, who have usually considered the design of a distribution system as something too complex for accurate computation, and yet too simple and practical to warrant such study. As a result, the art of designing transmission and distribution systems has developed slowly and has been out-distanced by the developments in supply, treatment, and conditioning of water. The distribution system is easily overlooked as it has usually grown bit by bit and seldom does the water-works man realize the accumulated investment that lies beneath the streets – out of sight and considered only when necessary."* Commenting on the importance given to the transmission and distribution mains in those days, Aldrich (1937) wrote that, *"in spite of the great sums of money invested in these facilities, no element of a water system receives less attention and design than the transmission and distribution system. Proper consideration given to the correct interpretation of the data upon which the design of these elements must be based, and to a study of the hydraulics involved, in conjunction with the field tests which are helpful in planning and extensions or improvements to the system, will go far in providing adequate and efficient installations with a minimum outlay of capital."* In those days the opportunity to design a complete water transmission and distribution system was seldom presented to water-works engineers. The water-works engineer's work was mainly concerned with providing extensions and reinforcements to existing systems. Hence the primary use for network analysis was in determining the necessity for, and the extent of, most effective reinforcements (Aldrich 1937).

Network Analysis Methods

Many methods have been used in the past to compute flows in networks of pipes. Such methods range from graphical methods to the use of physical analogies and finally to the use of mathematical models. Some of the methods were admittedly approximate while some of them were used for real time analysis and design. Each method possessed its own advantages and disadvantages and was best suited for a particular problem or application. Other methods were progressively modified to handle increasing complexities of the problem. A summary of some of the more important methods include:

- Freeman's Graphical Method
- The Equivalent Pipe Method
- The Electric Network Analyzer
- Hardy Cross Method (s)
- The Simultaneous Node Method
- The Simultaneous Loop Method
- The Linear Method (Simultaneous Pipe Method)
- The Gradient Method (Simultaneous Component Method)

The suitability, reliability, computational efficiency and accuracy of each of the above methods has been documented in the past and reported in the literature. Of the earlier methods, the equivalent pipe method, the graphical method, the electric network analyzer and the Hardy Cross method are the most important, not only as methods that were widely used in their respective periods but also because of the general impact they had on the field of pipe network analysis. The more recent methods such as the simultaneous node method, simultaneous loop method, the linear method, and the gradient method all use matrix formulations of the network problem in order to take advantage of the full power of modern day computers. Based on the methods reported in literature and the level of usage, the history of pipe network analysis may be divided into three distinct periods one from the late 1800s to the 1930s, one from the 1930s to the 1960s, and one from the 1960s to the present. Table 1 shows the chronology of the methods for pipe network analysis.

Graphical Methods

With the advent of the Darcy Weisbach equation in 1845, engineers were finally provided with a tool which could be used to predict headloss as a function of pipe discharge and the physical parameters of the pipe. However, due to the implicit relationship of the friction factor, the equation was largely restricted to applications to single pipe problems. One way to extend the applicability of the equation to water distributions systems was through the use of the graphical method. While Knapp (1937) noted that Spiess reported a graphical method for pipe network analysis in 1887 in "*Journal fur Gasbeleuchtung und Wassersorgung,*" the graphical method developed by Freeman (1892) appears to have dominated this period. Freeman demonstrated the application of his method by analyzing flow in compound pipes in series and parallel, and in a fairly complicated pipe network.

The graphical method involves preparing curves of discharge vs. head loss of pipes and complex arrangement of pipes. Starting from a simple pipe, a series of curves are developed step-by-step to include the complexities of pipes in series, pipes in parallel, pipes with take-outs (demands), pipes with put-ins (supplies), pipes with crossovers, compound storage, and pipe grid. In each of these curves the abscissa represents the rate of flow and ordinate the loss of head under the associated rate of flow. For example, for several pipes connected in series, curves are constructed for each pipe. The total loss of head for several pipe lines connected in series is obtained by adding the ordinates (headloss) of each section for each rate of flow. This is valid since the total head loss for a series of pipes is equivalent to the sum of the headlosses of each of the pipes. For the case of two pipes connected in parallel, the abscissas (flows) for each individual pipe which correspond to a common ordinate (headloss) are added together, since the loss of head through a given section consisting of several pipes connected in parallel is the same in each pipe and the total flow in the section is the sum of the flows in each pipe. The graphical solution of complex systems involving pipe arrangements having consumption at points along the line, the introduction of water from storage or from other sources, and those combinations of pipe for which several factors are

Table 1. Chronology of Pipe Network Analysis Methods by Periods

Year	Inventor(s) / Author(s)	Method / Application
Period I		
1845	Darcy and Weisbach	Formula for headloss for flow through single pipe
1892	Freeman	Graphical solution
1905	Hazen and Williams	Formula for headloss for flow through single pipe and Equivalent pipe method
Period II		
1934	Camp and Hazen	Electric network analyzer
1936	Cross	Relaxation technique
1956	McIlroy	McIlroy fluid analyzer
1957	Hoag and Weinberg	Adaptation of Hardy Cross method for digital computer
Period III		
1963	Martin and Peters	Simultaneous node method
1968	Shamir and Howard	Expansion of simultaneous node method
1970	Epp and Fowler	Simultaneous loop method
1977	Jeppson	Commercial network analyzing program based on simultaneous loop method
1972	Wood and Charles	Linear method
1980	Wood	KYPIPE – Commercial network analyzing program
1987	Todini and Pilati	Gradient method
1994	Rossman	EPANET - Commercial network analyzing program

dependent upon each other requires the graphical shifting of the discharge-head loss curves. This is done to represent the effect of variation in elevation of storage, the putting in or taking out of quantities of water, and the flow and pressure characteristics of centrifugal pumps. To facilitate the mechanics of the movement of these curves, convenient transparent media for the graphs were prepared in the form of small rectangles of about 4 in. by 6 in., cut from thin celluloid, processed or coated on one side to aid the drawing of the discharge-head loss curves and their removal when no longer required (Aldrich 1937). A final solution is obtained by "juggling" the position of the curves that represent the elements in the system.

Several documented cases of applications of the graphical method exist. In 1916 Kingsbury applied it for solving a composite twin pipeline problem, while in 1931, Palsgrove applied the method to a dual-flow water supply system (Gavet 1943). In 1934, Howland expanded Freeman's Graphical method to include take-off, and crossovers (Howland 1934). In 1937, Aldrich published a paper that contained a complete discussion of the application of the method to complicated networks with examples of solution. Aldrich reported that he had used the method for practical problems for more than ten years and had quite successfully checked his computations by the results obtained after the completion of several construction projects. Stanley (1937) claimed that the graphical method was well suited to problems with varying rates of flow, since the curves for different pipes and groups of pipes present a visual analysis of the happenings with varying flows. Direct mathematical computation, on the other hand, presents only the answer for a given condition and the effect of varying is not so easily visualized. Stanley also said that graphical method was exceptionally well adapted to problems involving storage, transmission mains, supply sources, pumping stations, and storage reservoirs or tanks since in such problems the graphical picture of the characteristics of the system permits visualization of hydraulic grade lines, friction losses. He also said that it might also be applied to small sections of a large system when it is desired to obtain a clearer "picture" of the relationship of flow, friction, and head varying conditions.

In addition to standard flow balance problems, the graphical method was also used to solve pumping problems. Aldrich (1937) claimed that he had used it for centrifugal pump studies, while Harold (1937) contended that several manufacturers of centrifugal pumps were using a method similar to this to show the behavior of pumps on a given pipe system. Harold also contends that the graphical method of analysis was very useful in investigating low-head complex pumping systems and easily solves some very difficult problems with remarkable accuracy. For an excellent review of the graphical method, as well as other early methods, the reader is referred to the paper by Gavett (1943) who noted at the time that: "*in a recent paper on the Hardy Cross method, the following statement was made: 'Prior to the inception of this method it was impossible to design, to any degree of accuracy, facilities for an increase of supply to any particular point in a grid system. Good judgment based on experience was the only basis of procedure.' The originators and proponents of the Hardy Cross method have never, to the writer's knowledge, made this claim. On the contrary, it was presented as a method that would give accurate results faster and easier than older methods. Its acceptance as a preferred method is attested by numerous papers by writers who have used it successfully. The discussion of older procedures is considered important, not only for historical accuracy, but also to call attention to alternate tools for doing the same work. Pipe flow studies are made for various purposes and the same method may not be best for all applications."*

The Equivalent Pipe Method

With the advent of the Hazen-Williams equation in 1905, engineers were provided with a way to easily decompose or aggregate composite systems of pipes into a hydraulically equivalent single pipe. This methodology became known as the Equivalent Pipe Method. The main application of the method has been in the design and analysis of transmission mains, although it has also been used to analyze aggregate flows over a water distribution system. In applying the method, a major path through the network is selected for analysis. In some cases, parallel lines are also included for analysis which are subsequently reduced to an hydraulically equivalent diameter, length, and/or roughness of pipe by application of an appropriate headloss equation (e.g. Hazen-Williams). In a similar manner, minor losses along the path are also typically accommodated by translation into an equivalent length of pipe. Ultimately, the entire path is decomposed into individual components which are then aggregated into a composite pipe of equivalent hydraulic characteristics. Once composed, the equivalent pipe can be used to determine the total headloss across the system under varying demand scenarios. Such calculations can also be readily used in sizing a pump for the system, or in evaluating the potential impact of a new parallel pipe.

The Electric Analyzer

In 1932 T.R. Camp, a Professor of Sanitary Engineering at MIT, and H.L. Hazen, a Professor of Electrical Engineering also at MIT, worked on an Electric Network Analyzer for pipe networks. A Network Analyzer developed for analyzing similar problems in distribution of electricity had been in use at MIT for some years then. The striking analogy between the basic laws governing hydraulic flow and electric current in networks suggested and encouraged the application of a similar method to pipe network analysis. In 1934 Camp and Hazen published their work on the electric network analyzer for analyzing flow in hydraulic networks. They applied the analyzer to solve a hydraulic network with several loops (Camp and Hazen 1934). As utilized in the solution of problems of hydraulic flow in pipes, the MIT network analyzer primarily is an assortment of resistors so arranged that they could be readily interconnected in a proper way to represent any desired pipe network. Pipes were represented by resistance units. Large-resistance units were used to represent pipes in which friction loss is great and small-resistance units were used to represent pipes in which friction loss is small. Units of each resistor were adjusted between wide limits by dial switches. Loads in hydraulic network were represented by large-resistance units. Interconnection of resistors is affected by means of a bus and flexible-lead scheme. This electrical "bus" corresponded to a hydraulic junction (Camp and Hazen 1934). Camp and Hazen claimed that the method was more precise than the hydraulic formulas that furnish the data for the problem. They also claimed, while talking about the need for such an analyzer, that the then existing methods for pipe grids were insufficient to apply to practical problems.

Even when the electric network analyzer produced reasonably good results, there was a serious problem in using the resistors to represent the pipes. This is because the voltage drop in a resistor is directly proportional to the current flow, but the head loss is an exponential function of discharge. Hence calibrating the resistors was a problem. To overcome this problem, several trial-and-error runs of the electric network analyzer had to be made. The need for a resistor element which possesses the same characteristics of the pipeline was clearly evident. As a result, beginning in 1934 studies were conducted at MIT under H.L. Hazen aimed at developing such an element. Camp (1943) reported several advancements pertaining to the electric network analyzers. According to Camp, in the early 1940s, Standrud developed a method for designing tungsten filaments satisfying the Williams-Hazen formula over a limited current range. In 1940, Laurent found that commercial vacuum tubes had the same property. Cook, in 1941, used commercial radio tubes. In 1942, Quill succeeded in efficiently solving the problem originally solved by Camp and H.L. Hazen in 1934. Collins and Jones solved the same problem in 1933 using an electric network analyzer with adjustable resistors (Camp 1943).

In the 1950s McIlroy developed the direct-reading electric network analyzer that used non-linear resistors (McIlroy 1950). This provided a rapid, convenient method to analyze water distribution systems as well as an easy way to visualize the results. Such analyzers were commercially manufactured and several installations were made. An excellent review of the McIlroy Analyzer was provided by McPherson and Radziul (1958). In 1956, Appleyard and Linaweaver introduced the McIlroy Fluid Analyzer, which used a fluister to represent a pipe element, and discussed its application to the cities of Baltimore and Philadelphia (Appleyard and Linaewaver 1956).

In general, the electric analyzer method generally produced good results once it was constructed and properly calibrated. However, the initial cost and expertise needed to set up the analyzer prevented its widespread use. Ironically, the development and initiation of this method had a secondary benefit: it revived interest in network analysis methods and ultimately influenced the development of the Hardy Cross method.

The Hardy Cross Method

In 1936, Hardy Cross, a structural engineering professor at the University of Illinois at Urbana/Champaign, developed a mathematical method for performing moment distribution analyses for statically determinate structures. Cross subsequently realized that the developed method could also be used to solve for pressures or flows in closed loop water distribution systems and published a paper outlining the application of the method for that purpose (Cross 1936). In that paper, Cross actually presented two different methods: one which solved for the flows in each pipe by the iterative application of a flow adjustment factor for each loop in the network, and one which solved for the hydraulic grades at each node in the system by an iterative application of a grade adjustment factor for each node in the system. In the former case, the associated nodal grades were then obtained by starting from a given reference point (e.g. a tank or reservoir) and then adding or subtracting the associated pipe headloss between adjacent nodes as determined by application of the Hazen-Williams equation. In the latter case, the flows in each pipe were then determined by solving the Hazen-Williams equation directly for discharge by using the differences in the resulting adjacent nodal grades as a measure of the headloss for the pipe. In applying the "loop" method, initial estimates of pipe flow had to be specified for each pipe which initially satisfied flow continuity at each junction node. In applying the "node" method, initial estimates of the nodal hydraulic grade had to be specified for each junction node.

In comparing the two methods, Cross noted that "convergence was slow and not very satisfactory" when employing the node adjustment method. This was attributed to the difficulty in obtained good initial estimates for the hydraulic grades. As a result, the loop adjustment method gained greater acceptance in the engineering community and quickly became known exclusively as the "Hardy Cross Method".

Many publications describing the application of the Hardy Cross method to various systems have appeared in the literature since Cross's original paper. In 1936 Donald applied the method to a slightly modified problem of one originally solved by Camp and Hazen in 1934 by the electric network analyzer (Donald 1936). In late 1939 Tyler applied the method to solve the problem of fire drafts of varying amounts from one or more centrally located fire hydrants in a symmetrical pipe grid system (Tyler 1939). He checked the accuracy of the graph giving head losses for grids of various pipe diameters which had appeared in the 1925 American Water Works Association Manual "Water Works Practice". He also checked the accuracy of the Circle method used for the hydrant analysis. He concluded that the curves in the AWWA manual were accurate and use of the circle method for hydrant studies is justified if the accuracy needed is low. However, Tyler advised the use of Hardy Cross method if higher accuracy was needed. In the early 1940s Hurst and Bubbis applied the method to a large distribution system (Hurst and Bubbis 1942). Others who applied the Hardy Cross method included Fair, Dodge, Hurst, Farnsworth Jr. and Rossano Jr., and Shand (Hurst and Bubbis 1942).

Early Reviews

One of the first critical reviews of methods for network analysis was provided by Camp (1943). Among the conclusions of his review were the following:

• The most widely used methods: Hardy Cross method, Electric network analyzer, and Freeman's graphical method are limited in their application to simple systems.
• All currently used methods of flow analysis then, including the Hardy Cross method, are not adequate for the design of any except very simple systems.
• The Hardy Cross method couldn't be applied to a large municipality without drastically skeletonizing the system.

Other authors besides Camp were also apprehensive about the economic feasibility and worthiness of physical modeling of distribution systems. Fair suggested it would be of greater benefit if that money were invested in field machinery such as main line meters, and gages and field personnel (Fair 1943). Engineers were also concerned about the effects of inaccuracies in fundamental assumptions that have to be made for coefficients of carrying capacity, rates of flow, and rates of consumption etc. Harold contended that practicality of models couldn't be fairly established until the amount of work and expense involved in reaching the required similitude was clearly realized (Harold 1943). He said the effect of incomplete field data, empirical formulae, and convergence of errors would affect the order of similitude between the model and prototype and suggested an error analysis method to estimate that similitude. Ways to adequately address many of such issues would have to wait the dawn of the computer age.

The Dawn of the Computer Age

In 1957, Hoag and Weinberg adapted the Hardy Cross method for solving the network flow problem to the digital computer and applied the method to the water distribution system of the city of Palo Alto, California. In presenting their application the authors discussed issues of speed, accuracy, and the economic viability of applying the method for a large class of network problems. Subsequently, two consulting engineering firms, Rader & Associates, Miami, and Brown & Caldwell, San Francisco, quickly emerged as early pioneers in the use of the computer to analyze water distribution systems for their associate clients (Eng. News-Record 1957). In that same year, an electronic computing firm, Datics Corporation of Fort Worth, Texas, became one of the first companies to sell a pipe network analysis computer program to one of its customers (Eng. News-Record 1957). The era of commercial software for network analysis had finally arrived.

Advanced Computer Methods

With the advent of increasingly sophisticated computers, more engineers began exploring the use of the Hardy Cross method to analyze flows and pressures in water distribution systems. However, with applications to larger systems, came a growing realization of some major limitations of the method: 1) depending upon the size and complexity of the system, the Hardy Cross method could sometimes take long periods to converge to a solution and in some instances failed to converge at all, and 2) the original method was restricted to closed loop systems and did not explicitly simulate the behavior of network components such as valves, pumps, etc. In response to such limitations, several researchers began investigating new mathematical formulations of the network analysis problem which could more fully take advantage of the opportunities afforded by high speed computations. Among the methods subsequently developed were 1) the simultaneous node method, 2) the simultaneous loop method, 3) the simultaneous pipe method, and 4) the composite method. A brief overview of each of these methods is provided below.

Martin and Peters (1963) were the first researchers to publish a computer algorithm that could be used to simultaneously solve for the hydraulic grades at each junction node in the distribution system. In essence, the method represented a simultaneous solution methodology for the original "node" method of Cross (1936). In applying the algorithm, the headloss equations for each pipe (e.g. Hazen Williams Equation) are written in terms of the flows in each pipe as expressed as a function of the hydraulic grades at the upstream and downstream ends of the pipe. Substitution of these equations into the associated conservation of mass equation for each junction yields a set of N (where N is the total number of junction nodes) nonlinear equations written in terms of the nodal grades. The resulting equations are then linearized using a standard Taylor Series expansion and solved iteratively using the Newton-Raphson method. Shamir and Howard (1968) demonstrated that the method could also be used to accommodate systems with pumps and valves and also showed how the method could be used to solve for other unknowns.

In 1970, Epp and Fowler applied the Newton-Raphson method to simultaneously solve for the flow adjustment factors associated with the original "loop" method of Cross (1936). This has the net benefit of significantly improving the convergence characteristics of the original algorithm. Subsequent researchers like Roland Jeppson at Utah State worked with CH2M Hill to develop a commercial program for network analysis based on the "simultaneous loop" method (Jeppson 1977). In using this approach, the nonlinear energy equations for each loop or path in the system are written in terms of flow adjustment factors. As with the "node" methods, the equations are linearized using a standard Taylor Series expansion and then solved iteratively using the Newton-Raphson method. Once the final adjustment factors are obtained, the individual pipe flows can be determined by multiplication of the original pipe flow estimates by the resulting factors. As with the original Hardy Cross method, the algorithm requires initial flow estimates for all pipes that satisfy flow continuity. In addition, like the original Hardy Cross method, determination of the associated nodal grades requires the subsequent application of the Hazen Williams equation for each pipe.

In 1972, Wood and Charles introduced yet another formulation of the network problem ("the linear method") in which the nodal conservation of mass and the conservation of energy equations for each loop or path are solved simultaneously to directly yield the flowrate in each pipe. As with the "simultaneous loop" method, determination of the associated nodal grades requires the application of a secondary headloss routine. However, by virtue of the combination of conservation of mass and conservation of energy equations, an initial flow balance of the nodes is no longer required. This method has the added advantage of being able to readily determine other unknown parameters besides flowrate. In some sense, the characterization of the method as the "linear method" is archaic and arose from the original way Wood and Charles proposed for minimizing the iterative convergence error associated with the solution of the nonlinear energy equations. Subsequent developments of the algorithm into commercial programs (i.e. WOODNET, KYPIPE, PIPE2000) actually employed a standard Newton-Raphson solution methodology (Wood 1980).

A final method for discussion is the Gradient Method, which was proposed by Todini and Pilati (1987). In this formulation, individual energy equations for each pipe are combined with the individual nodal equations for each junction node to provide for a simultaneous solution for both nodal heads and individual pipe flows. Similar to the "simultaneous loop" and the "linear method", the nonlinear energy equations are linearized using a Taylor Series expansion. However, in this case, the equations are solved using an efficient recursive scheme that employs an inversion of the original coefficient matrix. This method has been adopted for use by the EPA in the development of the program EPANET (Rossman 1994).

Summary

This article has presented a historical review of various methods for computing flows and pressures in water distribution networks from the late 19th century through the dawn of the 21st century. This era has witnessed the development of several innovative methods for network analysis, including such methods as the graphical method, the electric network analyzer, the Hardy Cross method, and the application of the Newton- Raphson method to various formulations of the conservation of mass and conservation of energy equations associated with water distribution networks. In addition to continuing to play a vital role in the design, operation, and management of water distribution systems today, such methods provide yet another example of the legacy of civil engineers and A.S.C.E. in impacting the quality of life in both the United States and the international community. With the advent of greater knowledge about the physical, chemical, and biological characteristics of water distribution systems, as well as the advent of new computer algorithms and associated computer technologies, the future opportunities for even greater impacts remain bright.

REFERENCES

- Aldrich, E.H. (1937). "Solution of transmission problems of a water system." Transactions of A.S.C.E., 1579-1619.
- Appleyard, V.A., and Linaweaver Jr., F.P. (1956). "The McIlroy fluid analyzer in water works practice." Journal of the American Water Works Association, 15-20.
- Camp, T.R., and H.L. Hazen (1934). "Hydraulic analysis of water distribution systems by means of an electric network analysis." Journal of the New England Water Works Association, 383-407.
- Camp, T.R. (1943). "Hydraulics of distribution systems – Some recent development in methods of analysis." Journal of the New England Water Works Association, 334-368.
- Cross, H. (1936). "Analysis of flow in networks of conduits or conductors." Engineering Experiment Station, University of Illinois, Bulletin No. 286.
- Donald, J.J. (1936). "Simplified analysis of flow in water distribution systems." Engineering News - Record, 117, 475-477.
- Epp, R. and Fowler, A.G. (1970). "Efficient code for steady-state flows in networks." Journal of the Hydraulics Division, A.S.C.E., 96, Proc. Paper 7002, 43-56.

• Fair, G.M. (1943) A discussion on "Hydraulics of distribution systems – Some recent development in methods of analysis," by T.R. Camp, Journal of the New England Water Works Association, 354-355.

• Field and Office (1957). "Computer firm sells pipeline net analysis." Engineering News-Record, 66.

• Freeman, J.R. (1892). "Arrangement of Hydrants and Water Pipes for Fire Protection." Journal of the New England Water Works Association, 7.

• Gavett, W. (1943). "Computation of flows in distribution systems." Journal of the American Water Works Association, 35(3), 267-287.

• Harold, E.B (1937). A discussion on "Solution of transmission problems of a water system." by Aldrich, E.H., Transactions of A.S.C.E., 1579-1619.

• Harold, T.M. Jr. (1943) A discussion on "Hydraulics of distribution systems – Some recent development in methods of analysis," by T.R. Camp, Journal of the New England Water Works Association, 355-368.

• Hoag, L.N., and Weinberg, G. (1957). "Pipeline network analysis by electronic digital computer." Journal of the American Water Works Association, 517-524.

• Howland, W.E. (1934). "Expansion of the Freeman method for the solution of pipe flow problems." Journal of the New England Water Works Association, 408-411.

• Hurst, W.D., and Bubbis, N.S. (1942). "Application of the Hardy Cross method to the analysis of a large distribution system." Jl. American Water Works Association, 34, 247-271.

• Jeppson, R.W. (1977). "Analysis of flow in pipe networks." Ann Arbor Science, Ann Arbor, MI, 1977.

• Jesperson, K. (2001). "A brief history of drinking water distribution." On Tap, 18-46.

• Knapp, F. (1937). A discussion on "Solution of transmission problems of a water system." by Aldrich, E.H., Transactions of A.S.C.E., 1579-1619.

• Martin, D.W. and Peters, G. (1963). "The application of Newton's method to network analysis by digital computer." Journal of the Institute of Water Engineers, 17, 115-129.

• McIlroy, M.S. (1950). "Direct-reading electric network analyzer for pipeline networks." Journal of the American Water Works Association, 347-365.

• McPherson, M.B., and Radziul, J.V. (1958). "Water distribution design and the McIlroy network analyzer." Journal of the Hydraulics Division – Proceedings of the A.S.C.E., 1588-1 to1588-9.

• Rossman, L.A., Clark, R.M. and Grayman, W.M. (1994). "Modeling chlorine residuals in drinking-water distribution systems." Journal of Environmental Engineering, A.S.C.E., 120(4), 803-820.

• Shamir, U. and Howard, C.D.D. (1968). "Water distribution systems analysis." Journal of the Hydraulics division, A.S.C.E., 94, Proc. Paper 5758, 219-234.

• Stanley, C.M. (1937). A discussion on "Solution of transmission problems of a water system."by Aldrich, E.H., Transactions of A.S.C.E., 1579-1619.

• Todini, E. and Pilati, S. (1987). "A gradient method for the analysis of pipe networks." International Conference on Computer Applications for Water Supply and Distribution, Leicester Polytechnic, UK.

- Tyler, R.G. (1939). "Flow of water in pipe grid systems." Journal of Water Works and Sewerage, 86(8), 285-288.
- Wood, D.J and Charles, C.O.A. (1972). "Hydraulic network analysis using linear theory." Journal of the Hydraulics division, A.S.C.E., 98, Proc. Paper 9031, 1157-1170.
- Wood, D.J. (1980). User's Manual - Computer Analysis of Flow in Pipe Networks Including Extended Period Simulations, Department of Civil Engineering, University of Kentucky, Lexington, KY.

The Bureau of Reclamation's Legacy: A Century of Water for the West

Elaine Simonson[1] and Toni Rae Linenberger[2]

Abstract

The Bureau of Reclamation, a bureau within the U.S. Department of the Interior, was established as the Reclamation Service in 1902. Its purpose was to settle the West by providing essential water supplies. Reclamation has accomplished that mission and now seeks to manage, develop, and protect water and related resources in an environmentally and economically sound manner in the interest of the American public.

Reclamation is celebrating 100 years of providing water to the western United States. It began its centennial on June 17, 2002, with an event at Hoover Dam, one of the bureau's most well-known, largest facilities. Approximately 2,200 people attended the event. Other events and exhibits will occur throughout the West during the centennial year.

Reclamation's history program is an essential part of the bureau, providing an oral history program, a project history program, and other preservation tools. Much of this information is available to those interested in learning about the bureau's history.

The Past

The Bureau of Reclamation's history and western U.S. water development are closely linked: you cannot get a clear perception of one without understanding the other.

---

[1]Public Affairs Officer, Washington Public Affairs Office, Bureau of Reclamation, P.O. Box 25007, D-1540, Denver, CO 80225-0007; phone 303-445-3139; esimonson@do.usbr.gov.

[2]Historian, Office of Policy, Bureau of Reclamation, P.O. Box 25007, D-5300, Denver, CO 80225-0007; phone 303-4452912; tlinenberger@do.usbr.gov.

Harnessing water through irrigation was essential for successful settlement of the West's arid lands. When elected, Theodore Roosevelt stated: "While I had lived in the West I had come to realize the vital need of irrigation to the country and . . . The first work I took up when I became President was the work of reclamation."[3]

With backing from Roosevelt and other key supporters, the Reclamation Act was passed in June 1902. It authorized the Secretary of the Interior to locate and construct irrigation projects in 16 of the 17 arid states and territories west of the Mississippi River. (Texas was subsequently added in 1906.)

During its early history, many viewed Reclamation as a positive, powerful force.[4] The bureau fulfilled Roosevelt's "homemaking" vision, supplying essential water supplies that built the western economy. Many of the West's great cities, including Boise, Salt Lake City, Spokane, and Phoenix, were built around Reclamation projects. The bureau supplied invaluable food and power supplies during World War II. It also employed thousands of Americans during the Great Depression.

Although many Reclamation projects were originally built for a single purpose—primarily irrigation—additional benefits such as flood control, recreation, municipal water supplies, and hydroelectricity quickly became vital to westerners.

The Present

Reclamation's mission has evolved throughout history to reflect the changing needs and values of the American public. With the exception of the Animas la Plata Project being constructed in southwestern Colorado and northwestern New Mexico, the bureau is no longer building large water storage facilities. Instead, it focuses on responsible water management to maintain its facilities, meet water delivery obligations and explore innovative methods to stretch existing water supplies while balancing competing uses.

Even after 100 years, Reclamation's impact on the lives and livelihoods of westerners remains undiminished:

- Reclamation is the nation's largest wholesale water supplier, bringing life-giving water to one-third of the irrigated agricultural lands in the West. One out of every five western farmers receives water to irrigate farmland, producing 60 percent of the nation's vegetables and 25 percent of its fruits and nuts. Reclamation also supplies drinking water to more than 31 million people.

- Reclamation is the second largest hydropower producer and the ninth largest utility in the U.S. The bureau's 58 hydroelectric powerplants supply about one-third of the power generated in the West, producing enough electricity to serve nine million people, generating over $600 million in annual revenues.

---

[3] Theodore Roosevelt, *An Autobiography.* 1913. McMillan Press, New York City, New York.
[4] William E. Warne, *The Bureau of Reclamation.* 1973. Praeger Publishers, Inc. Boulder, Colorado.

- More than 90 million people visit Reclamation's 340 reservoirs each year to enjoy boating, fishing, hiking, hunting, and other recreational activities.

- Water from Reclamation projects provides significant environmental benefits, creating wetlands, fish and wildlife habitat, and water quality enhancement while satisfying original project purposes.

- To date, Reclamation personnel have provided technical assistance in water management to more than 80 countries and training to more than 10,000 international colleagues.

The Future

The need for safe, adequate water is proving to be one of this century's critical dilemmas. John W. Keys, III, Reclamation's Commissioner, envisions Reclamation as playing a key role in addressing and resolving the complex water management issues the West faces today.

"Reclamation is working in collaboration with the states, our partners, and all water users to leverage resources and make efficient use of developed water supplies," he states. "We must continue to seek for innovative strategies where water can be used more than once and, thus, satisfy multiple entities."[5]

Keys also believes in turning challenges into opportunities through the use of what Interior Secretary Gail Norton's calls the "4Cs"---communication, consultation, and cooperation—all to serve the cause of conservation and maintenance of our limited water resources.

Bureau priorities include the following:

- Operating and maintaining projects in a safe and reliable manner to protect the health and safety of the public and employees.

- Building partnerships with broad groups of stakeholders and interest groups to meet the increasing demands for limited water resources.

- Assisting states develop contingency plans for droughts.

- Supporting water districts, cities, and others to develop and implement water conservation and recycling measures.

---

[5] John W. Keys, III. March 2002. Commissioner, Bureau of Reclamation, 1849 C Street NW, Washington, DC 20240-0001

- Developing and implementing innovative methods, such as water banking and desalination, to increase available water supplies.

- Continuing to provide sustainable hydropower supplies and meet water delivery obligations.

Preserving Reclamation's History

Reclamation's history program, established in 1988, serves to collect and preserve the bureau's history. Beginning in about 2001, the program focused on Reclamation's centennial in 2002.

The yearlong observance of the centennial began with an event at Hoover Dam on June 17, 2002. Roughly 2,200 invited guests – including Secretary of the Interior Gale Norton; Reclamation Commissioner John Keys, III; several former Secretaries of the Interior and Reclamation Commissioners; dignitaries; employees; retirees; and employee and retiree family members – enjoyed the event. Attendees listened to Secretary Norton and Commissioner Keys laud the accomplishments of the bureau during the last 100 years. The evening was capped off with a fireworks display and laser show.

Other centennial-related items include three traveling trailers that will visit different locations in the seventeen western states throughout the year, as well as a traveling exhibit and an art and photograph exhibit.

A history symposium was held on June 18 and 19, 2002, in conjunction with the University of Nevada, Las Vegas. This event brought together scholars, students, and interested parties for a discussion of Reclamation's history and its role in the West. The forty-five papers presented during the two days will be available sometime this fall or winter.

The major centennial publication, a history of Reclamation's hydropower program entitled *Dams, Dynamos, and Development: The Bureau of Reclamation's Power Program and Electrification of the West*, debuted at the celebration on June 17. A full color, large-format book, the history covers the evolution of hydropower within Reclamation and places this history into the wider context of western development. The publication has been distributed to Reclamation employees and retirees. Copies can be obtained from the Superintendent of Documents at the Government Printing Office.[6]

In addition to its focus on centennial activities, the history program continues ongoing activities, including gathering oral histories and writing project histories.

Collecting oral histories that retain the corporate memory of bureau personnel is a major component of the history program. To date, numerous interviews with a wide variety of

---

[6] United States Department of Commence, National Technical Information Service. 5285 Port Royal Road, Springfield, Virginia 22161; phone1-800-553-6847; www.ntis.gov.

current and former specialists and personnel have been completed; in some cases, spouses have also contributed. Most of the living former commissioners have participated in this program. With approval from interviewees, histories are available to employees, researchers and other interested parties.

The program includes extensive oral histories on the Newlands Project near Fallon, Nevada. One historian interviewed many of the community members associated with the project to gain a better understanding of a functioning Reclamation project as a whole. The project was chosen for this in-depth study because it is small and has a wide variety of issues – including legal, water rights, environmental protection, Native American water rights, and groundwater.

In 2002, Reclamation contracted with Professor William Rowley of the University of Nevada, Reno, a highly respected, independent scholar, to write an unbiased history of the Bureau of Reclamation. Plans are to publish Professor Rowley's book some time in 2005 or 2006.

The history program also preserves histories for Reclamation's individual projects. In fact, brief project histories have been written for many of the 180-plus projects. When completed, the histories will be edited and published in a single document. These histories can be found at **www.usbr.gov** and several major libraries, including the Western History Department at the Denver Public Library, the Newberry Library in Chicago, the Water Resources Archives at the University of California at Berkeley, the Huntington Library in San Marino, California, the Beinecke Library at Yale, the American Heritage Center of the University of Wyoming, and the Congressional Research Service.

The history program staff is available to help researchers – both internal and external.

For more information please contact:

Brit Storey, Senior Historian
PO Box 25007 (D5300)
Denver, CO 80225-0007
303-445-2918
bstorey@do.usbr.gov

Toni Rae Linenberger
PO Box 25007 (D5300)
Denver, CO 80225-0007
303-445-2912
tlinenberger@do.usbr.gov

Impact of Reclamation's Hydraulic Laboratory on Water Development

Philip H. Burgi [1]

ABSTRACT

The paper covers the history of Reclamation's hydraulic laboratory from its inception in 1930 at Colorado Agricultural Experiment Station, Fort Collins, Colorado to the present. Emphasis is placed on the laboratory's historical role in developing new design concepts for hydraulic structures to meet Reclamation's ever-increasing challenges over the past seventy years.

The paper presents the design challenges associated with specific structures such as: Hoover Dam side channel spillway, Grand Coulee Dam spillway bucket, Hungry Horse Dam tunnel spillway, and more recently the aeration slot design developed for Reclamation's tunnel spillways to prevent cavitation damage.

During the 1950's and 1960's Reclamation's hydraulic laboratory initiated an extensive research program to develop standard designs that eventually led to engineering monographs and manuals coauthored by hydraulic laboratory staff. The paper concludes with the hydraulic design challenges facing Reclamation in the next century. The water management issues associated with fish passage and water conservation as well as infrastructure security at numerous dams in the western U.S. are some of the hydraulic challenges in Reclamation's future.

BACKGROUND

The Bureau of Reclamation was established in 1902. In its first ten years eighteen dams were built. By 1930 fifty dams had been constructed. The first irrigation projects were fairly simple, consisting of a diversion dam, headworks, canals, and turnouts. These early projects involved no special challenges other than those peculiar to each site. To optimize water basin development, dams of increasing height were required and their design and construction created new problems and provided serious challenges for Reclamation's engineers.

The 1906 Congress introduced the function of hydropower when it authorized the sale of excess power generated at Reclamation projects. In 1928 Congress passed the Boulder Canyon Project Act (The name Boulder Dam was changed to Hoover Dam April 3, 1947 by joint congressional resolution). This act inaugurated a new era in the conservation and utilization of western water. Hoover Dam would be the principal structure of the Boulder Canyon Project and would introduce a new concept in western water development referred to as multi-purpose development. Other projects soon followed: the Central Valley Project, Columbia Basin Project, Colorado-Big Thompson Project, and the Missouri River Project. These multi-purpose projects optimized utilization of water and land resources in large areas of entire river basins. Rhone[1] states the quarter century between 1948 and 1973 was especially productive when more than half of Reclamation's dams were constructed.

THE EARLY YEARS

In the early years before 1930, many of Reclamation's design engineers were recruited from Reclamation's parent organization, the U.S. Geological Survey. The

---

[1] Founding Member EWRI of ASCE, Water Resources Consultant, Wheat Ridge, Colorado,

supervisory staff of the design units maintained extremely high engineering standards for their personnel. Typically, each design leader assembled and maintained a design manual based on their training and experience; these informal manuals were passed on to subordinates who, in turn, added to the standards and through their new knowledge and experience became even better qualified designers.

When Reclamation completed the construction of Shoshone Dam (100 m) in Wyoming in 1910, it was the highest dam in the world. In the next 25 years Reclamation held this record three more times with the construction of Arrowrock Dam (106 m) in Idaho in 1915, Owyhee Dam (127 m) built in eastern Oregon in 1933, and finally Hoover Dam (221 m) on the Colorado River in 1936.

RECLAMATION'S HYDRAULIC LABORATORY

Reclamation's hydraulic laboratory was established in the early 1930s expressly to solve the technical challenges presented in the design of these large structures. With the anticipation of designing Hoover Dam there came the recognition that this structure would impose design and construction challenges well beyond the textbooks and experience of the day. The tremendous construction costs associated with these large structures required careful attention to the preliminary design and required hydraulic model testing before one could finalize design and start construction.

Although the name "hydraulic laboratory" is relatively modern, the concept has been around for a long time. Scholars as early as Leonardo da Vinci recognized the importance of experimentation when dealing with the flow of water. He is quoted as saying, "Remember when discoursing on the flow of water to adduce first experience and then reason". [2]   The purpose of the hydraulic model is to use the tool of similitude to demonstrate the behavior of flowing water at reduced scale. Typically, models are used to study rivers and waterways of hydraulic structures and equipment such as: spillways, outlet works, stilling basins, gates, valves, and pipes associated with large dams. Agreement between model and prototype has proven very satisfactory. [3, 4]

At the turn of the 20th century, some European universities and especially universities in Germany recognized the value of experimental model studies to solve hydraulic challenges such as those posed by dam spillways and outlet works, siphons, tunnel inlets, and bridge constrictions on rivers. John R. Freeman (1855-1932), a hydraulic engineer from the United States, felt very strongly that we should develop similar hydraulic laboratories to those being utilized in Europe. In 1924 he visited laboratories in Berlin, Dresden, Brunn and Karlsruhe. He had a significant influence on the development of hydraulic laboratories in the United States. Freeman writes in 1929, *"Nowhere, yet, in America has the writer found the acceptance and reliance upon the doctrines of similitude which he has found at substantially all of the great European engineering universities, and which have been developed there wholly during the past 30 years, and mainly during the past 10 years."* [5]

Beginning in the early thirties, laboratory activity in engineering schools in the United States greatly increased. Freeman describes some of the early work conducted in laboratories in the United States: Cornell University (1899), State University of Iowa (1919), Alden hydraulic laboratory of the Worcester Polytechnic Institute (1910), and several commercial laboratories conducting experimentation with hydraulic turbines.   Eventually, hydraulic laboratories were established in government facilities such as the Miami Conservancy District in Ohio, the U.S. Bureau of Standards, U.S. Army Corps of Engineers, Soil Conservation Service, and the U.S. Bureau of Reclamation.

Figure 1.   1931 Photo of Reclamation Hydraulic Laboratory staff at Fort Collins

Investigations with hydraulic models had their start in the Bureau of Reclamation in August 1930 when thirteen engineers, technicians, and craftsmen from the Denver Reclamation Office began working in the hydraulic laboratory of the Colorado Agricultural Experiment Station in Fort Collins, Colorado. The 242 m^2 laboratory was built in 1912 under the direction of Ralph Parshall.

By 1935, the laboratory in Fort Collins had expanded to four times its original size to handle the ever-increasing Reclamation work load.   One of the early studies was for the proposed shaft spillways for Hoover Dam.   As a result of these studies a change was made from the original shaft spillway concept to two side-channel spillways to accommodate the design flow that had increased from 5,670 to 11,340 m^3/s.

In the summer of 1929, Emory Lane was appointed as engineer in charge of the Bureau of Reclamation's of hydraulic, sediment, and earth materials research studies.   A graduate of Purdue and Cornell Universities, he worked for the Miami Conservancy District, Ohio before coming to Reclamation.   During his 6-year period as administrator of the hydraulic laboratory, Lane initiated the comprehensive laboratory investigations undertaken for Hoover Dam, Grand Coulee Dam, Imperial Dam and de-silting works and the model studies of the All American Canal structures.

Jacob Warnock, another graduate of Purdue, came to Reclamation as an associate hydraulic engineer after working with the Corps of Engineers in their Chattanooga, Nashville, and Huntington offices.   By 1934 Warnock, became head of the hydraulic laboratory in Fort Collins when Emory Lane moved to Denver to direct a small hydraulic laboratory that had been set up in the basement of the Old Custom House in Denver. Victor Streeter, who later became a renowned Professor of Hydraulics at the University of Michigan, was one of the staff members in Denver during this period.

Figure 2: Jacob Warnock with visiting engineers in custom House Laboratory

In a summary article written in 1936, Warnock stated, *"Models were first used extensively by the Bureau in 1930 in the design of the spillway for the Cle Elum Dam of the Yakima project in Washington. The design of the spillways for Boulder Dam, Madden Dam in the Panama Canal Zone, and Norris and Wheeler dams for the Tennessee Valley Authority, served as stepping stones in further developing the technique and improving the methods."*[6]

Warnock was a strong believer in the value of hydraulic model investigations. *"The procedure by which models of hydraulic structures are built and tested in the laboratory before the design is finally adopted and committed to construction is analogous to the manner in which a newly designed machine is thoroughly inspected for defects and imperfections at the factory. The models reveal undesirable features of the design and indicate the proper means for the correction."*[7]

By 1935 Jacob Warnock became head of the laboratory in Denver and was instrumental in its move to the New Custom House in 1937 where there was approximately 475 m^2 available for studies.

The work of the laboratory became so prolific that Reclamation tested 80 models in the period from 1930-38 and had 50 engineers, technicians, and craftsmen working in three laboratories. *"The use of models has proved so advantageous in indicating opportunities for reducing costs and improving hydraulic properties that the work of the laboratories is now recognized as a regular part of hydraulic design. At the present time, the three laboratories are engaged in testing or constructing models of twenty different features relating to ten major projects."* [7]

In the fall of 1938, Reclamation discontinued its work in the Fort Collins laboratory. Warnock figured prominently in the design of the hydraulic features of Hoover, Grand Coulee, Shasta, Friant, and many other large dams and irrigation projects in the west. His untimely death in December 1949 at the age of 46 was a great shock to Reclamation's Denver Center.

The wartime westward shifting of population and industry created an impetus and need for a Reclamation construction program much larger after the war than it had been before. By 1943 Reclamation organized into seven regional areas based on large watersheds in the West and established a Chief Engineer's Office in Denver responsible for all design and construction. The small laboratory space in the New Custom House was inadequate for the enlarged program. Sufficient space was available at the former Denver Ordnance Plant (Remington Small Arms Plant) located on the west side of Denver and now referred to as the Denver Federal Center (DFC).

In the later part of 1946, the hydraulic laboratory was moved to its present home in the Denver Federal Center where it occupied some 4925 m^2 of laboratory space. At the time, Reclamation's staff at the Denver Federal Center totaled over 2240 employees. These facilities were unequaled in their specialized qualifications anywhere in the world. Design and construction engineers worked in tandem with experts in hydraulics, concrete, soils, chemical, and other laboratories to meet the new challenges of water development in the arid west.

A quote from the July 1950 edition of The Reclamation Era states, *"The combination of men and laboratory equipment is paying huge dividends to the public. Water and power users, who ultimately pay for Reclamation projects, pay for the work of the Branch of Design and Construction. They should be reassured to know that economies in construction discovered at the Center have more than paid for its total operating costs, as well as the entire cost of establishing and equipping it. Many of the money-saving techniques and materials conceived in connection with specific construction works will apply as well to later works, thus compounding the monetary economies."* [8]

There were other hydraulic laboratories developed and used by Reclamation. They were primarily field laboratories located at: Montrose, CO (1931-1936), Grand Coulee Dam (early 1940s), Hoover Dam (1939-1945), Estes Park Colorado Powerplant (late 60s and early 70s).

LABORATORY CONTRIBUTIONS

Spillways

Spillways at dams are used to pass the design flood and thus protect the dam from overtopping. Early in Reclamation history there were five general categories of spillways in use: "glory hole" or shaft-type (Gibson Dam), side-channel (Hoover Dam), overflow type (Grand Coulee), open chute type (Bartlett Dam), and enclosed tunnel chute (Seminole Dam).

The importance of adequate spillway design cannot be overemphasized.

Operating experience with spillways for dams has revealed problems of two types: (1. inadequate capacity, and (2. unsatisfactory performance for design or less-than-design discharges. Historically, Reclamation has taken a very serious position toward adequately studying spillway performance before going to final design.

One of the first major impacts resulting from hydraulic laboratory studies was the major improvement in spillway capacity resulting from the replacement of the planned glory-hole spillway design for Hoover Dam spillways with the side-channel spillway that ultimately provided the desired spillway capacity.

These early model studies were conducted at Ft Collins and Montrose as well as the Custom House in Denver. The large 1:20 scale outdoor model at Montrose was used to finalize the design of the drum gates on the side-channel spillways at Hoover Dam (total spillway capacity of 11,340 m^3/s) replacing the proposed Stony gates, which proved to be unsatisfactory during the model tests. A total of eight models were used in the hydraulic design of Hoover Dam with model scales of 1:20(2), 1:60(3), 1:64, 1:100, and 1:106.[9]

Four models were used in the design of Grand Coulee Dam ranging in scale from 1:30 to 1:184. A major improvement in the design for Grand Coulee Dam was the replacement of a proposed large hydraulic jump stilling basin with a roller bucket to dissipate the energy at the toe of the Grand Coulee spillway designed to pass 28,325 m^3/s. A construction savings of $4,750,000 (1941 costs) resulted from use of the roller bucket energy dissipator developed in Reclamation's hydraulic laboratory. [8]

Reclamation's high dam tunnel spillways proved to be a very economical means to pass large flood discharges in lieu of building large capacity surface spillways and stilling basins on the dam abutments. However, as early as the winter of 1941 when the Arizona tunnel spillway at Hoover Dam operated for 116 days there was suspicion of the vulnerability of concrete to damage caused by high velocity flow in tunnel spillways.[10] This spillway operation resulted in a large hole in the tunnel spillway elbow 14 m deep, 9 m wide and 35 m long. The damage was thought to initiate at a "misalignment" of the tunnel invert just above the elbow. The damage was caused by high velocity flow passing over the roughness and leading to bubble formation (similar to boiling water) in the flow. When the bubbles collapsed, high energy shock waves were generated damaging the concrete. This phenomena is referred to as cavitation formation and damage. In the 1940's the damage was repaired by backfilling with river rock and then covering with a thick layer of high quality concrete. The concrete surface had a very fine finish, almost terrazzo, to prevent reoccurrence of the cavitation. Tunnel spillways were later constructed at Yellowtail, Flaming Gorge, Blue Mesa, and Glen Canyon Dams.

The cavitation damage problem surfaced again in June and July 1967 when the tunnel spillway at Yellowtail Dam discharged for 20 days at 425 m^3/s. By July 14 it was evident that there was a problem in the tunnel spillway. When drained and inspected a hole 2 m deep, 6 m wide and 14 m long was discovered. In earlier laboratory investigations, the introduction of as little as 7.5% air into the water flow eliminated damage associated with cavitation on concrete surfaces.[11] In 1967, Hydraulic laboratory studies on a 1: 49.5 scale model of the Yellowtail Dam tunnel spillway resulted in design of an aerator located some distance upstream of the elbow consisting of a 760 mm high ramp that extended above the springline of the tunnel and provided air to the underside of the high velocity jet traveling through the tunnel.

The first installation of an aerator in a tunnel spillway was at Reclamation's Yellowtail Dam.[12]

In 1983, high runoff in the Colorado River basin created the need to pass flood flows through tunnel spillways at Blue Mesa, Flaming Gorge, Glen Canyon, and Hoover Dams. The resulting damage was so extensive at Glen Canyon Dam's two tunnel spillways that $42,000,000 (1985 costs) and a year of reconstruction was required to repair the spillways and install an aerator in each tunnel.[13] Reclamation conducted extensive laboratory model tests to determine hydraulic performance of the aerators at these tunnel spillways.

Figure 3. Damage to Glen Canyon Dam left spillway in 1983

By 1985 aerators were installed in all five of these high head tunnel spillways in the western United States. The left tunnel spillway (Arizona side) at Hoover Dam experienced cavitation damage in 1983 and had to be repaired with an aerator added despite the smooth surface placed in 1943. Henry Falvey wrote a comprehensive engineering monograph summarizing Reclamation's experiences and developments in cavitation damage control entitled, *Cavitation in Chutes and Spillways*.[14] This publication was yet another of the numerous documents produced by the hydraulic laboratory staff to assist in the design of water projects.

Sediment Control Structures at Diversion Dams

In the period from 1950-1965 numerous model studies were used to develop sediment control measures at diversion dams.[15] To develop the most satisfactory solution of a sediment control problem at a diversion usually requires a "movable bed" hydraulic model study. Structures and techniques such as curved guide vanes, short tunnel under sluices, and vortex tubes were developed in the laboratory to exclude sediments. On large projects such as the All-American Canal, large settling basins were developed and built. However, the cost of these large structures was prohibitive for many of the diversion dams across the Plains States. More economical solutions were often developed which included a simple gated sluiceway and using some of the water as a means to bypass the sediments around the diversion intake structure.

Gates and Valves

It was clear in the 40's that as the size of dams and reservoirs increased, for economic reasons it became necessary to design projects for multiple use, such as flood control, irrigation, power development, and river regulation for navigation. The rigorous demands imposed by such multiple use of a storage dam required that the outlets be designed to give close regulation of the rate at which stored waters were released. The increase in dam height lead to higher pressures and velocities and in many cases the need for larger capacity outlets. Many improvements in the mechanical design of gates and valves were made to meet the challenge of these new conditions. However, most gates and valves were designed for simple open or closed operations. Regulation in some cases was made by providing numerous outlets controlled by gates such as those used at Grand Coulee Dam where increase or decreases could be made in finite increments equal in value to the capacity of a single outlet. Most valves developed prior to the 1930's were designed for pressure heads up to 130 m, totally inadequate for the new dams proposed.

The Hoover Dam tunnel-plug outlets provided the most outstanding challenges. Each tunnel had six – 1830 mm needle valves under pressure heads up to 171 m which discharge up to 623 m^3/s into a 15 m diameter concrete tunnel. The laboratory model studies included tests at scales of 1:106, 1:60, and 1:20 to assure the validity of the design against any scale effects. The final configuration selected represented a distinct improvement over those originally proposed. The laboratory tests also showed that large air vent tunnels originally proposed were not necessary resulting in construction savings of $30,000 (1932 costs).[16] There were several occasions in the 1980's where the old internal differential needle valves failed during uncontrolled closure. In some cases, these uncontrolled closures resulted in loss of life. In the early 1990's Reclamation undertook additional studies to replace all of their needle valves across the West. The needle valves were soon replaced with large jet flow gates developed by Reclamation in the late 40's for Shasta Dam.

The preferred large valves for Reclamation dams were the needle valves (1909-1942) and the hollow jet valves (1950-1967). Over the years, Reclamation has upgraded outlet gates and valves from the early Ensign valves (1905-1915), to needle valves (1909-1942), to tube valves (1941-1945), to hollow-jet valves (1950-1967) and jet-flow gates (1945-67). James Ball, Donald Colgate and Donald Hebert were three key hydraulic laboratory contributors to Reclamation's work in the development of high-head outlet gates.[17] A 1973 American Society of Civil Engineering article gives a summary of some of these gates and valves and their installations across the United States.[18]

Hydraulic Laboratory Techniques

In 1955 hydraulic laboratory personnel published Engineering Monograph No.18 entitled *Hydraulic Laboratory Practice*. It was prepared as an aid in applying engineering knowledge and experience to hydraulic laboratory studies. Emphasis was placed on the basic principles of similitude; techniques of model design, construction, and operation; equipment; and field studies. The volume which has been used in hydraulic laboratories world-wide was updated in 1980 on the golden anniversary of the Bureau of Reclamation's first hydraulic model tests.[19]

Stilling Basins and Energy Dissipators

Although hundreds of stilling basin and energy dissipating devices have been designed and built for spillways, outlet works, and canal structures, it is often necessary to make model studies of individual structures to be certain that these will operate as anticipated. In the early 1950's a ten-year laboratory research effort was undertaken to develop general design criteria for stilling basins and energy dissipators. Existing information was gathered from laboratory and field tests collected from Reclamation records and experiences over a 23-year period. Hundreds of additional tests were conducted using six laboratory test flumes. The largest flume was 102 mm wide, 24 m long with an available height of 5.5 m and a discharge capacity of 800 l^3/s. Tests included hydraulic jump stilling basins, short stilling basins for canal structures and small spillways, wave suppressors for canal structures, sloping apron stilling basins, slotted and solid bucket energy dissipators, baffled apron drops, tunnel spillway flip buckets, and test to size riprap downstream of stilling basins. This effort conducted solely in Reclamation's hydraulic laboratory and supervised by Alvin Peterka, resulted in the world renowned Engineering Monograph No. 25 entitled, *Hydraulic Design of Stilling Basins and Energy Dissipators* [20], which has been used for many years as a standard for such hydraulic structures world-wide.

By the 1970s the trend for spillway terminal structures had returned to the flip bucket - the principle used was to direct the flow away from the structure and downstream a sufficient distance where the water could erode its own plunge pool or discharge into a pre-excavated plunge pool. Devices such as a combined hydraulic jump/flip bucket were used for the tunnel spillway at Yellowtail Dam and the surface spillway at McPhee Dam. The energy is dissipated within the basin at the end of the tunnel spillway up to a predetermined discharge where the jump flips out and the structure acts as a flip bucket for larger discharges. Most of the tunnel spillways previously mentioned terminate with flip buckets designed based on various hydraulic model studies in the 50's and 60's.

A device called a baffled apron drop was developed in the laboratory primarily for use on canals as a drop structure at wasteways. In the late 1970's laboratory staff started looking at the baffled apron drop as a spillway structure for dams. In the 80's many baffled apron drops were used as spillways on several Reclamation dams as well as for the States of Washington, New Mexico and Nevada (Conconully Dam, Truth or Consequences Dam, Marble Bluff Dam).

THE HYDRAULIC LABORATORY IN THE 21st CENTURY

At the beginning of the 21st century, Reclamation continues to use the laboratory facilities at the DFC, however there have been many changes over the past 70 years. There are new and improved microprocessor laboratory controls. An ozonator system has been installed to improve water quality and provide for longer use of recirculated water. There have been giant strides in electronic control and measurement as well as increased use of hybrid modeling where numerical and physical modeling techniques are brought together to better understand fluid mechanics. Skilled craftsmen who build the intricate models have always been part of the laboratory staff and continue to play a key role in laboratory studies.

Reclamation's hydraulic structures and equipment investigations and development in the period from 1930 through the 1970's resulted in world class technological advancements in water-resource development. However by the latter quarter of the 20th century, a major paradigm shift had occurred with water development in the United Sates.   As public values shifted toward more environmental sensitivity, water agencies changed their focus from an emphasis on *water development* to *water management*.   Reclamation's hydraulic laboratory program maintained a contemporary focus throughout these changes over time. The new focus led to an emphasis on developing improved technologies for *(1. protecting the public and existing water infrastructure, (2. encouraging water-use efficiency,* and *(3. emphasizing environmental enhancement on regulated river systems.* [21]

In the area of <u>water infrastructure protection,</u> Reclamation's hydraulic laboratory has played a key role in the development of cost-effective spillway designs focused on dam safety issues. Alternative spillway designs, fuse plug concepts, and overtopping protection concepts have been tested and developed.   Laboratory research on the labyrinth spillway concept produced design criteria that were applied to the 14 cycle labyrinth spillway for Ute Dam in New Mexico.[22] The labyrinth spillway resulted in construction savings of over $24,000,000 (1982 costs) compared to a traditional gated structure at Ute Dam.[21]

Another alternative spillway design gaining acceptance in the engineering community is the fuse plug concept. Reclamation's hydraulic laboratory advanced the science and acceptance of fuse plugs now used at several Reclamation dams.[23] The construction savings realized by using fuse plugs for additional spillway capacity for Horseshoe and Bartlett Dams on the Verde River in Arizona were in the range of $150-300 million (1984 costs).[21]

Stepped spillway design criteria developed in Reclamation's hydraulic laboratory played a pivotal role in its world-wide acceptance in the 1990's. Stepped spillways are very compatible with Roller Compacted Concrete (RCC) construction and provide an economical spillway when constructed as an integral part of the dam. Hydraulic model studies of stepped spillways for McClure, Milltown Hill, Stagecoach, and Upper Stillwater Dams in the late 1980's were critical in defining energy dissipation characteristics and hydraulic performance of this new concept.[24]

Another recent advancement has been the protection of embankment dams during overtopping occurrences.   Studies performed in Reclamation's hydraulic laboratory as well as tests performed in a large-scale outdoor overtopping facility at Colorado Sate University have proven the viability of 305 mm wide, 51 mm high, and 610 mm long concrete blocks to protect the surface of an embankment. [25]

<u>Water-use Efficiency</u> continues to play an important role in Reclamation's program. The Western United States depends on a water storage and delivery system built over the past 150 years to provide water for irrigated agriculture, municipal and industrial use, power generation, and recreation. Population growth and environmental water requirements place additional demands on a limited supply and require managers to look for water-use efficiencies. In response to this reality, the hydraulic laboratory has placed increased emphasis on conservation technologies. The ability to measure discharge in open channels on Reclamation projects has been dramatically improved in the last twenty-five years by the development, in

cooperation with Agricultural Research Service, of the long-throated flume and broad-crested weir measurement methods as well as other technologies that are robust, low cost and accurate.

In 1953 Reclamation's hydraulic laboratory produced the first edition of the *Water Measurement Manual*. It was compiled from Reclamation's Manual for Measurement of Irrigation Water published in 1946. A second edition was published in 1967. The most recent edition published in 1997 still emphasizes the basics of water measurement but is updated to include the latest measurement technologies.[26] It is also available at: **http://www.usbr.gov/wrrl/fmt/wmm/index.htm**

In addition to water measurement, the laboratory staff has worked for over thirty years in development of water system automation technologies. Reduced cost and increased capabilities of sensors, computer hardware, software, and data telemetry systems have brought practical canal automation capabilities within reach of the majority of water and irrigation districts in the western United States, including many smaller and older districts that still operate their systems using the same methods used decades ago.

Future water development will be closely linked with *environmental enhancement* as Reclamation continues to play a role in providing a high standard of living while protecting environmental resources. Historically, Reclamation has had a concern for the natural environment especially as it may impact fish and wildlife resources. In the late 1950's Reclamation's hydraulic laboratory staff assisted with the development and field and laboratory testing of a pilot fish screen structure constructed in the headworks of the Tracy Pumping Plant.[27]

More recently, several fishery and stream restoration projects have built on this earlier experience and illustrate this new enhancement approach. To improve the winter-run Chinook salmon population in the Sacramento River, the laboratory initiated an aggressive research study to develop temperature-control curtains in reservoirs such as Lewiston and Whiskeytown Lakes.[28] The use of this new temperature-control technology, as well as the steel shutter structure at Shasta Dam, has increased the selective withdrawal capability within the Sacramento River basin and improved the management of the river temperature by several degrees and greatly improving the habitat for anadromous fish species. The laboratory has also been involved in retrofitting several Reclamation dams to provide selective withdrawal capability: Shasta, Lewiston, Whiskeytown, Hungry Horse, and Flaming Gorge Dams.

Within Reclamation a bioengineering focus (biological science and engineering) has led to new, innovative concepts for using hydraulic structures to manage regulated water systems in the West. This cooperation of hydraulic engineering and biological sciences in recent years has produced innovative technologies for fish screening, fish separation and handling, and fish passage upstream and downstream at dams and diversion works. These research efforts and experiences will soon be published as a Reclamation fisheries manual. On many Reclamation projects these advancements have been crucial to maintaining water deliveries while also providing new environmental benefits.

Summary

The history of Reclamation's hydraulic laboratory is a story of engineers, technicians and craftsmen who have had an attitude and work ethic best characterized by their persistent high quality work used to tackle the challenges of water development in the West. To some degree, they were exceptional individuals but for the most part their greatest achievements resulted from their ability to work as a team. Although some individuals have been mentioned in this paper, one needs to recognize that the greater gains were almost always the effort of a team. There are many excellent engineers on the present staff who no doubt will become part of the great legacy of Reclamation's hydraulic laboratory. Future generations will make those judgments. Suffice to say, that Reclamation and the nation have benefited greatly by the productivity of the hydraulic laboratory staff over the past seventy years. There are new challenges facing today's laboratory engineers and scientists and their responses to these challenges will define the future legacy of the laboratory.

References:

1. Thomas J. Rhone, 50th Anniversary of the Hydraulics Division 1938-1988, ed. Adnan M. Alsaffar (ASCE, 1990), 132-147.

2. Gunther Garbrecht, ed., Hydraulics and Hydraulic Research A Historical Review-International Association for Hydraulic Research 1935-1985, A.A.Balkema/Rotterdam/Boston/1987.

3. "Conformity Between Model and Prototype in Hydraulic Structures", Transactions, American Society of Civil Engineers, Vol 109, pg 3-193.

4. Philip H. Burgi, ed, Model-Prototype Correlation of Hydraulic Structures, Proceedings of the International Symposium, American Society of Civil Engineers, August 9-11, 1988. Pp 1- 480.

5. John R. Freeman, ed., Hydraulic Laboratory Practice, The American Society of Mechanical Engineers, 1929, 1-4, 17, 28-32, 697-742.

6. Jacob E. Warnock, "Experiments Aid in Design at Grand Coulee", Civil Engineering, November 1936, pp.737-741.

7. U.S. Department of the Interior, Bureau of Reclamation, Dams and Control Works, February 1938, 205-212.

8. U.S. Department of the Interior, Bureau of Reclamation, The Reclamation Era, February, 1941, 29-32 and July 1950,127-128.

9. Boulder Canyon Project Final Reports, U.S. Department of the Interior, Bureau of Reclamation, Part VI- Hydraulic Investigations, Bulletins 1-3, 1938.

10. Keener, "A Challenge to Hydraulic Designers", Engineering News Record, November 18, 1943.

11. A. J. Peterka, "The Effect of Entrained Air on Cavitation Pitting", Proceedings of

the Joint Meeting of the International Association for Hydraulic Research, ASCE, Minneapolis, MN, August 1953.

12. D. Colgate, "Hydraulic Model Studies of Aeration Devices for Yellowtail Dam Spillway Tunnel", U.S. Department of the Interior, Bureau of Reclamation, REC-ERC-71-47, 1971, 13.

13. P.H. Burgi and M.S. Eckley, "Repairs at Glen Canyon Dam," Concrete International: Design and Construction of the American Concrete Institute, March, 1987, pp. 24-31.

14. H.T. Falvey, Engineering Monograph No. 42 Cavitation in Chutes and Spillways, U.S. Department of the Interior, Bureau of Reclamation, 1990, 145.

15. Enos J. Carlson, "Laboratory and Field Investigations of Sediment Control Structures at Diversion Dams", Ninth Congress on Irrigation and Drainage, International Commission on Irrigation and Drainage, Moscow, 1975.

16. Engineering News Record, "Hoover Dam Number", December 13, 1932.

17. D.J. Hebert and J. W. Ball, "The Development of High-Head Outlet Valves", Report on Second Meeting, Appendix 14, International Association for Hydraulic Structures Research, Stockholm, Sweden, June 6-7, 1948.

18. Cortland C. Lanning, "High Head Gates and Valves in the United States", Journal of the Hydraulics Division, ASCE, Vol. 99, No Hy 10, October, 1973, pp. 1727-1775.

19. J.C. Schuster, ed., "Hydraulic Laboratory Techniques", U.S. Department of the Interior, Bureau of Reclamation, 1980, 208.

20. A. J. Peterka, Engineering Monograph No.25 Hydraulic Design of Stilling Basins and Energy Dissipators, U.S. Department of the Interior, Bureau of Reclamation, 1964, 217.

21. Philip Burgi, Issues and Directions in Hydraulics , eds. Tatsuaki Nakato and Robert Ettema A.A.Balkema/Rotterdam/Brookfield/1996,187-198.

22. K.L. Houston, "Hydraulic Model Study of Ute Dam labyrinth Spillway", U.S. Department of the Interior, Bureau of Reclamation, GR-82-7, August, 1982, pp 41.

23. C.A. Pugh, "Hydraulic Model Studies of Fuse Plug Embankments", REC-ERC-85-7, December, 1985.

24. Kathleen L. Houston and Alan T. Richardson, "Energy Dissipation Characteristics of a Stepped Spillway for an RCC Dam", International Symposium on Hydraulics for High Dams, Beijing, China, November 15-18, 1988.

25. Kathleen H. Frizell and Brent W. Mefford, "Designing Spillways to Prevent Cavitation Damage", Concrete International: Design and Construction of the American Concrete Institute, May, 1991, pp. 58-64.

26. Water Measurement Manual,3rd ed., U.S. Department of the Interior, Bureau of Reclamation, 1997, pp.400.

27. Thomas J. Rhone and Daniel W. Bates, "Fish Protective Facilities at the Tracy pumping Plant-Central Valley Project California", 9th Hydraulics Division Conference, American Society of Civil Engineers, Seattle, WA, Aug 17-19, 1960.

28. Tracy B. Vermeyen and Perry L. Johnson, "Hydraulic performance of a flexible curtain used for selective withdrawal- a physical model and prototype comparison", Proceedings, ASCE National Hydraulic Engineering Conference, 1993.

High Dams and Large Reservoirs; The Evolution of Concrete Dams at the Bureau of Reclamation

Gregg A. Scott[1], P.E., M. ASCE, Larry K. Nuss[1], P.E., John H. LaBoon[1], P.E.

Abstract

Over the past 100 years the Bureau of Reclamation has made significant contributions to the advancement and evolution of concrete dam design, analysis, and construction. This paper chronicles some of those achievements.

Introduction

As the Bureau of Reclamation (Reclamation) celebrates its centennial (1902-2002), it is a good time to look back at the engineering accomplishments of the organization. This is nowhere more evident than in the advancements made in the area of concrete dam design, analysis, and construction. Reclamation has designed and constructed over 50 major concrete storage dams throughout the Western United States. This paper discusses some of the contributions made by Reclamation along the way, beginning with the early masonry dams, and proceeding through to applying the technology to dam safety modifications.

The Early Years

Reclamation's long history of concrete dam design and construction began in September 1903 at a conference of Reclamation Service engineers in Ogden, Utah. It was recognized there that Reclamation would be required to build masonry dams of great height in order to store the water required to reclaim the arid lands of the west. This lead to the unique design and construction of Pathfinder Dam on the North Platte River in Central Wyoming (Wisner, 1905). Due to the narrow granite canyon in which the dam was to be constructed, it was decided that an arch should be built, and that the arch action could be relied upon. It was recognized that masonry dams were far from rigid, and that temperature was an important problem. The modulus of elasticity and coefficient of thermal expansion were estimated for a composite of rock blocks and concrete. Using these properties, the dam was designed as a combination of horizontal arches and a vertical crown cantilever. The load was distributed between the arch and cantilever so as to produce equal deflections at their point of

---

[1] Bureau of Reclamation, P.O. Box 25007, Denver, CO 80225

intersection. The stresses from the deflections were then calculated, and the cross-section proportioned accordingly. This was the early beginnings of what was later to become known as the "Trial Load" method of analysis.

Although facilitated by steam power, many of the construction techniques adopted for the construction of Pathfinder Dam are still in use today. A large tunnel was constructed to divert the flow of the river, and was later used for the outlet works. A key component to the foundation excavation and dam placement was an overhead cableway. An aggregate crushing plant and concrete batch plant were constructed on site. It was recognized that an impervious dam could be built at the same cost as a leaky dam, the main difference being more rigid inspection and an understanding at the start that only first-class work would be allowed. This became the practice for all Reclamation concrete dams. All foundation rock and material to be placed in the dam was thoroughly washed and cleaned. Granite stone from the spillway excavation was set and vibrated into place between masonry courses on the upstream and downstream face, with concrete and smaller rocks worked into the joints to produce a well-consolidated solid mass. Construction of the dam began in 1906, and was completed in 1909 to become Reclamation's first masonry dam. The dam is 65.2 m (214 ft) high, stores a reservoir of about $1.23 \times 10^9 \text{ m}^3$ (1,000,000 acre-ft).

Completed in 1910, a similar cross-section and arch shape was adopted for the highest dam in the world at the time (99.1m, 325 ft) across the Shoshone River in Northwestern Wyoming. Originally called Shoshone Dam, it was later to become known as Buffalo Bill Dam. Concrete was placed between wooden forms, as shown in Figure 1. The concrete was placed in eight-inch layers, and granite plum rocks, comprising about 25 percent of the concrete volume, were placed in the concrete and shaken or rammed into final position, solidifying the mass to a considerable degree. Spading and tamping was performed to work the concrete into all the cavities and ensure consolidation against the forms. The plum rocks projected above the cleaned lift surface for bonding to the next layer.

Figure 1. Concrete Placement at Buffalo Bill Dam

The first use of radial contraction joints occurred at East Park Dam in Northern California, as shown in Figure 2. The radial joints were spaced at 6.1 m (20ft), and a shear key 0.15 m (6 in) deep by 0.61 m (2 ft) long was constructed in the joints about six feet from the upstream face. Although there is no indication that waterstops were installed in the joints, a system of 102-mm (4-in) diameter tile drains was constructed downstream of the keys to convey water from the joints to the outlet tunnel. This dam was also constructed entirely of concrete. The original design called for sandstone blocks to be imbedded in the concrete to make up 20 to 30 percent of the mass. However, the sandstone was of poorer quality than first believed, and the sandstone blocks were omitted from the construction. A conservative gravity arch design was adopted for this structure, completed in 1910, at 42.7 m (140 ft) high.

Figure 2. Construction of East Park Dam

The reign of Shoshone Dam as the world's highest was short-lived. In 1916, Arrowrock Dam was completed by Reclamation in Idaho, to a height of 106.4 m (349ft) (Bureau of Reclamation, 1954). The gravity arch design made use of radial contraction joints. Three vertical wells were formed in each joint, which were later filled with concrete during cold weather, after the dam had undergone contraction. A Z-strip annealed-copper waterstop was installed in each joint five feet from the upstream face. Immediately downstream of this strip a triangular drain was formed, to collect water and transport it to galleries constructed within the dam. The concepts of vertical formed drains within the concrete, and foundation drainage and grouting appear at Arrowrock Dam. "Sand cement", containing an equal amount of Portland cement and pulverized sand, reground to such fineness that 90 percent would pass a No. 200 sieve, was used in constructing the dam. Although this saved on the quantity of cement used, the concrete did not attain as much strength, and as a result the durability suffered. This concept was therefore abandoned.

The Amazing Arch and Developments of the 1920's

During the 1920's materials were relatively expensive, and there was a desire to minimize the amount of concrete used in constructing dams. Independent arch theory became the order of the day for designing concrete dams in narrow canyons, where the thickness of the arch at any given elevation was designed independently. Thinner dams resulted from this method of design.

In 1925 Reclamation completed Gerber Dam, a 25.9-m (85-ft) high thin arch dam on Miller Creek in Southern Oregon. The concrete was placed in 1.2-m (4-ft) lifts between keyed contraction joints spaced at 15.2-m (50-ft) centers. To facilitate cooling and contraction of the dam, closure slots were left on each side of the central overflow spillway section, as shown in Figure 3. Concrete was placed in these slots at low temperature conditions, once the dam had cooled. Extensive field testing was used for quality control, including sieve tests of the sand and aggregate, compression tests on concrete cylinders of concrete taken from the forms, and slump tests. A slump up to three inches was permitted to allow the concrete to flow through the placement chutes. Gerber Dam represents the first use of instrumentation in a Reclamation concrete dam. Electric resistance thermometers, Berry strain gages, and survey targets were installed in the dam. Despite the success of the thin arch dams of the 1920's, Reclamation would opt for thicker and more massive structures throughout the next few decades.

Figure 3. Gerber Dam Under Construction

Prelude to Hoover Dam

Hoover Dam was on the drawing boards in the late 1920's, and it was recognized that with the great size of this dam would come additional problems. In preparation for this enormous project, important developments occurred during the design and construction of Owyhee Dam, which was completed in 1932, immediately preceding Hoover Dam. At 127.1 m (417 ft) high, this gravity-arch structure was another world's highest. Independent arch analysis was abandoned in favor of a gravity arch section analyzed by the trial load method, which advanced throughout the 1920's.

The materials and construction techniques were similar to earlier structures, but a few items were notably different. These largely related to the temperatures generated by the hydrating cement, and the possibility for cracking if not handled properly. Large eight- to nine-inch cobble rock aggregate was added to the concrete mix to reduce the cement per unit volume of concrete. A system of pipes was installed along the vertical contraction joints, spaced at 15.2 m (50 ft), to cool the concrete mass. Cement grout was then forced into the joints under pressure through the piping system, to lock in the arch action when the concrete was at its coolest point. Grout zones were 30.5 m (100 ft) high, and isolated with 20-gage soft copper sheets. Experimental cooling systems were installed at various locations, whereby cooling coils were placed on the top of concrete lifts and cold water circulated through the system. Measurements were taken to determine the effectiveness of these systems to reduce the heat and thermal gradients in the concrete, and to open the contraction joints for grouting. The systems were effective, and the information was put to good use during the design and construction of Hoover Dam.

Hoover Dam - Quantum Leaps Forward

There is no question that Hoover Dam was the crowning achievement that came to represent the Bureau of Reclamation's world renowned expertise in concrete dam design and construction (Figure 4). At 221.6 m (727 ft) high, Hoover Dam was the highest dam in the world for quite some time. It is still the second highest dam, and the highest concrete dam in the United States. When designs for Hoover Dam began, it was to be more than double the height of Arrowrock Dam, then the highest dam in the world. As such, it was evident from the start that many new problems in design and construction would require solution before the dam could be built. As a result of intensive research, improvements were made in practically every aspect of the design of the dam, waterways, and appurtenant structures. The work is published in a series of reports, which represents some of the most comprehensive and finest documentation of its kind (Bureau of Reclamation, 1938-1948). It would not be possible to do justice to this body of work in a short paper such as this.

Extensive use was made of analytical and physical models in the design of Hoover Dam. Thermal analyses were developed based on data obtained from Owyhee Dam. The trial load method was brought to maturity and used for extensive analyses of the dam. These calculations were checked by the use of two- and three-dimensional physical models. Two-dimensional slab analogy experiments of a cantilever and arch were used to obtain stress functions usable in the trial load analyses. Stresses in the slab are proportional to twist and curvature in the slab under the applied boundary loading. Three-dimensional model studies showed stress concentrations at the top of the dam where there was a change in arch length. As a result, fillets were added to increase the thickness of the dam near the abutments.

Extensive research was conducted into the cement and concrete mix to be used in the dam. Large concrete cylinders, 0.9 m (3 ft) in diameter by 1.8 m (6 ft) high, were cast and tested to examine the effects of the 229-mm (9-in) maximum size aggregate. One of the major problems was controlling the temperatures that would result from the rapid construction and unprecedented size of the dam, locking in temperatures that would take more than 100 years to dissipate on their own. As a result, low heat cement was developed for most of the construction. Still, the mass concrete needed to be artificially cooled by circulating cold water through cooling pipes placed at the top of each 1.5-m (5-ft) high lift of the 15.2-m (50-ft) by 15.2-m (50-ft) construction blocks. A 2.4-m (8-ft) wide slot was left open down the middle of the dam for the extensive system of cooling pipes. The cooling slot was filled and the contraction joints grouted once the dam had cooled. Due to the size of the dam, large plants had to be erected to process aggregate and sand, and to batch the concrete in large quantities. The 2,485,000 m^3 (3,250,000 yd^3) of concrete were placed from June 1933 to May 1935, in 23 months.

Figure 4. Hoover Dam after 15 Months of Concrete Placement

Hydraulics for High Dams

The unprecedented size of the spillways and hydraulic head at Hoover Dam without question resulted in a major breakthrough in spillway design. In particular, research and development of methods to design the "ogee" spillway crest is still used for spillway designs around the world (Brandley, 1952). A second major breakthrough occurred with the first printing of Engineering Monograph No. 25 in 1958 (Peterka, 1984). This publication summarized 23 years of research and design experience and provided a generalized and practical design tool for sizing stilling basins and other energy dissipating structures. Prior to this time, attempts to generalize data from hydraulic model studies lead to inconsistent results. A third major advancement in evaluating hydraulics for high dams involved an understanding of cavitation. Cavitation is the result of formation and collapse of vapor cavities at abrupt changes in geometry of the flow surface, which can result in severe erosion and damage to concrete and rock. The large flow velocities associated with high dam waterways makes them more susceptible to cavitation. Reclamation had investigated cavitation damage and implemented repairs as early as 1941. However, it was not until much later that Engineering Monograph 42 was published, providing common-sense guidance on how to identify and mitigate cavitation potential (Falvey, 1990).

World War II - Large Gravity Dams

The United States entered World War II in 1941, and electric power was needed to fuel war production factories. The design and construction technology developed at Hoover Dam was put to work in completing large multi-purpose dams with associated powerplants. Many of these were constructed on large and wide rivers, where it became necessary to construct concrete gravity dams. The largest, and perhaps best known, is Grand Coulee Dam (Figure 5) in Washington State, which has been studied and emulated around the world (Bureau of Reclamation, 1953). As finally constructed, this structure is 550 feet high and nearly a mile long. The central spillway section is capable of passing a flow of 28,300 m^3/s (1,000,000 ft^3/s). Three-dimensional trial load twist analyses, fully developed during the design of Hoover Dam, were performed for the high gravity dam design. Due to stress concentrations in the portion of the dam adjacent to the sharply rising abutments and concerns for potential cracking, vertical "twist slots" were constructed in the abutment sections to provide flexibility and allow the dam to adjust to the rising reservoir levels. The slots were initially filled with sand, and after the reservoir had filled, the sand was removed and the slots filled with concrete. By this time Reclamation's 22,200 KN (5,000,000lb) testing machine was installed in the Denver laboratory at the U.S. Customs House, and was used in testing the strength of the large aggregate concrete used in the dam, using cylinders 0.9 m (36 in) in diameter. Many of the same construction techniques used at Hoover Dam, including the use of low heat cement and cooling coils, were again used at Grand Coulee. Two batch plants were constructed, one on each side of the canyon. At the peak of production, 15,814 m^3 (20,684 yd^3) of concrete were placed in a single day. With the completion of the Forebay Dam and Third Powerplant in 1974, the total volume of concrete in the dam is well over 7,600,000m^3 (10 million yd^3), and the total power generating capacity is over 6000 MW.

Figure 5. Grand Coulee Dam Under Construction

The Post-War Boom – Developments Continue

Following World War II, the country entered into a boom period. During this period, problems that developed with concrete durability at some earlier structures were solved. Concrete dams in cold climates, particularly those constructed with wet concrete mixes, suffered from freeze-thaw deterioration. This was addressed by the use of air-entraining admixtures in the concrete mixes. Alkali-aggregate reaction occurred in some structures, where the concrete aggregates chemically reacted with the alkali in the cement, resulting in expansion and cracking of the concrete. The use of low alkali cement in the concrete mix was found to alleviate this problem. Extensive instrumentation systems became standard for measuring the response of the structures. Thermal analysis became standard (Townsend, 1981). These advancements, and the technology developed from building large concrete dams and powerplants was quickly put to use in building several more monumental concrete arch dam structures. Hungry Horse Dam is a 172-m (564-ft) high structure on the South Fork of the Flathead River in Montana; Glen Canyon Dam is a 216-m (710-ft) high structure on the Colorado River upstream of the Grand Canyon; Yellowtail Dam is a 160-m (525-ft) high structure on the Bighorn River in Montana; and Flaming Gorge is a 153-m (502-ft) high structure on the Green River in Utah.

Rock Mechanics and Foundation Design Emerge

The 1959 abutment sliding failure of Malpasset Dam in Southern France made the profession recognize the need for more rigorous foundation investigations and analytical design methods. Shortly after this, Reclamation began developing rock mechanics methods in application to concrete dam foundation design and analysis. However, it was not until the foundation designs were underway for Auburn Dam in the late 1960's that the foundation exploration, analyses, and design were coherently integrated. Although Auburn Dam was never completed, this work was an enormous contribution to the profession, and formed the basis for future evaluations within the Bureau of Reclamation (Bureau of Reclamation, 1977-1978).

The Auburn Damsite consists of complex metamorphic geology. Careful diamond core drilling using split inner tube core barrels, trenching, and excavation of exploratory tunnels and drifts was performed to define the geologic conditions. The results were portrayed on geologic plan, section, and structural contour maps to provide a complete three-dimensional picture of the foundation and the discontinuities within the rock. Weathering profiles were developed and fracturing was characterized to determine appropriate excavation depths. Methods were developed to extrapolate results from large-scale in-situ deformation testing to develop deformation properties for the entire foundation. From this, the deformation properties of the foundation were defined for input to finite element and trial load structural analyses of the concrete arch dam. Potential modes of instability were identified and analyzed by examining discontinuities (faults, shears, joints, foliation planes, etc.) within the foundation. This "failure mode assessment" as it is sometimes called was developed fully in the rock mechanics arena, and has been a valuable contribution to other areas of engineering. Based on this testing and analysis, foundation treatment was designed to mitigate any area of concern, be it controlled by deformation, seepage characteristics, or stability of the rock mass.

The Double-Curvature Arch – A New Standard for Efficiency

Beginning in the 1960's a new concept for shaping arch dams found its way to the Bureau of Reclamation (Boggs, 1975). This shape, termed "double curvature", provided for more efficient distribution of loads within the structure. A double-curvature arch is curved in section view as well as plan view. This results in somewhat of a "bowl" shape. The undercutting at the heel of the dam and the inward curvature on the downstream face that results from this shape eliminate areas where tensile stresses typically develop in arch dams.

The first double-curvature dam constructed by the Bureau of Reclamation is Morrow Point Dam. At a height of 143 m (468 ft), the dam is located in a narrow section of the Black Canyon of the Gunnison River in Colorado. The "trial load" method of analysis provided information that was used in shaping the dam. Several other double-curvature arch dams were successfully designed and constructed by the Bureau of Reclamation in the 1960's and 1970's. One that bears mention is Nambe Falls Dam, a 45.7-m (150-ft) high structure on Rio Nambe in New Mexico. The arch is part of a composite structure, with a massive concrete thrust block on the left abutment that ties into an embankment dam. It was difficult to design for the temperature conditions at the site. Therefore, a series of flat jacks were installed in the crown cantilever joint, and the flat jacks were pressurized to prestress the dam into a state of compression that could handle all loading conditions adequately. Reclamation also pioneered the development of elliptical arches by the use of "three-centered" geometry. The elliptical arches are approximated by a central section with a smaller radius, flanked by abutment sections with larger radii. This allows double-curvature arch dams to be designed for wider canyons. Although none of these were built by Reclamation, the method was developed and several designs were completed.

Figure 6. Upstream Face of Morrow Point Dam during Construction

Structural Analysis Developments

Prior to the availability of digital computers in the 1960's, trial load analyses were performed by engineers operating mechanical "adding machines" and filling in values on large tables. They would work in pairs, and one would check the other's computations as they were performed. One analysis would take a pair of engineers six to eight weeks. As such, not many load combinations were analyzed. The seating arrangement in the Analysis Unit was like a Viking ship with the rowmaster (Unit

Head) behind the rowers (Engineers). As computers became available, procedures were developed to solve the "trial load" problem directly, and the computer program ADSAS (Arch Dam Stress Analysis System) was developed for this purpose.

In the 1970's the country became more aware of the potential hazard associated with earthquake loading. Although ADSAS had some limited seismic analysis capability, it lacked the ability to perform modern response analyses. During the design of Auburn Dam, the finite element method was used to perform linear elastic, three-dimensional, response history modal superposition analyses (Bureau of Reclamation, 1977-1978). Since that time, every concrete storage dam in Reclamation's inventory has received similar analysis. Evaluating the results of dynamic finite element analyses required advances in estimating concrete strengths under seismic loading. Rapid loading laboratory tests were developed and performed which confirmed that an increase in tensile strength of about 50 percent can be expected under dynamic loading.

Continued advances in the finite element and finite difference techniques have accompanied advances in computer capabilities. Frequency domain solutions were developed at the University of California, Berkeley, for seismic analysis incorporating coupled hydrodynamic and foundation interaction. Nonlinear finite element codes allow modeling of contraction joints and nonlinear concrete material behavior. Although it is more difficult to interpret the results from such analyses, they more realistically represent dam behavior, and are capable of predicting actual failure modes. Stress and load distribution can change dramatically when geometric and material nonlinearities are incorporated. Some recent dam safety evaluations have incorporated these programs and capabilities.

Roller-Compacted Concrete – Rapid Construction for Gravity Dams

The concept of roller-compacted concrete (RCC) involves placement of a lean and dry concrete mix by spreading it in thin layers with a bulldozer, and compacting it with vibratory drum rollers. The lean mix reduces the generated heat, and rapid production rates can be achieved using a mechanized construction process. Reclamation began testing a high paste (cement plus flyash) RCC mix in 1980. This resulted in a strong and stiff material with similar properties to conventional concrete, and allowed gravity dams of this material to be designed using conventional methods. In 1985, placements began at Upper Stillwater Dam, Reclamation's first RCC dam and at the time the world's largest. The straight gravity dam is about 85.3 m (280 ft) high, 823 m (2700 ft) long, and contains over 1,220,000 m^3 (1,600,000 yd^3) of concrete. Laser-guided slip forms were used to place concrete elements forming the upstream and downstream face of the dam. RCC was placed between the facing elements. The RCC contained more than twice as much flyash as cement. In 1986, over 547,000 m^3 (715,000 yd^3) of RCC was placed in less than five months, and the peak shift placed over 4100 m^3 (5400 yd^3). The major drawback to the design and construction of Upper Stillwater Dam was the exclusion contraction joints. This resulted in formation of vertical cracks through the structure, and significant leakage, requiring remedial measures. Methods were subsequently developed to include joints and waterstops in RCC structures.

Transitioning to Dam Safety – Applying Technology to Reduce Risk

As construction of new concrete dams was winding down, renewed emphasis was placed on dam safety due to an Executive Order issued in April 1977. The aim of Reclamation's Dam Safety Program is to ensure that the Agency's dams do not pose an unacceptable risk to the downstream public. Reclamation has used concrete dam design, analysis, and construction technology developed over the past century to meet this objective. The most recent involved stabilizing Pueblo Dam on the Arkansas River in Southern Colorado. Nearly horizontal foundation shale layers beneath the massive head buttresses of the dam daylighted in the large stilling basin excavated at the toe of the dam. The reservoir had never fully filled. Due to the large population downstream of this dam, potential sliding of the structure on these shale layers posed a high risk. A RCC plug and toeblock anchored with double corrosion protected high strength rock bolts were constructed in the stilling basin to block the potential sliding planes (Figure 5). State-of-the-art distinct element and probabilistic stability analyses were performed to ensure the RCC geometry would be effective in stabilizing the dam. Contraction joints were formed in the RCC by vibrating steel plates into the freshly compacted lifts. The contraction joints in the cross-canyon direction needed to be grouted to ensure that load could be transferred from upstream to downstream with minimal displacement. Six-inch diameter holes were drilled along the joints in locations where the steel plates had been omitted. Tubing was designed and installed in the holes to provide grout supply and return lines, and venting to remove air and water from the system. Grouting of the joints was performed the second winter following RCC placement when joint meters indicated sufficient joint opening for the grouting operations. The grouting was successful, and the joints did not close the following summer, indicating good filling of the joints.

Figure 7. RCC Placement at Pueblo Dam

Final Thoughts

We hope you have enjoyed this tour of the evolution of concrete dam design, analysis, and construction within the Bureau of Reclamation over the past century.

There is no question that the early pioneers in this effort were extremely talented and set the stage for some of the great feats of human engineering that were to follow. Monumental projects like Hoover and Grand Coulee Dams are still "wonders" today. During the heyday of dam construction in the United States, the Bureau of Reclamation developed a reputation as a world leader in concrete dam technology. The construction of large dams in the United States is winding down now after a century of extensive development. The Bureau of Reclamation can be proud of the legacy, expertise, and technology it has left the Nation and the world in concrete dam design and construction.

References

Boggs, H.L. (1975). *Guide for Preliminary Design of Arch Dams*, Engineering Monograph No. 36, Second Printing, U.S. Department of the Interior, Bureau of Reclamation, Government Printing Office, Washington, DC.

Brandley, J.N. (1952). *Discharge Coefficients for Irregular Overfall Spillways*, Engineering Monograph No. 9, U.S. Department of the Interior, Bureau of Reclamation, Denver, Colorado.

Bureau of Reclamation (1938-1948). *Boulder Canyon Project Final Reports*, U.S. Department of the Interior, Bureau of Reclamation, U.S. Government Printing Office, Washington, D.C.

Bureau of Reclamation (1954). *Dams and Control Works*, Department of the Interior, Bureau of Reclamation, Government Printing Office, Washington D.C., Third Edition.

Bureau of Reclamation (1977-1978). *Design and Analysis of Auburn Dam*, U.S. Department of the Interior, Bureau of Reclamation, Denver, Colorado.

Falvey, H.T. (1990). *Cavitation in Chutes and Spillways*, Engineering Monograph No. 42, Eighth Printing, U.S. Department of the Interior, Bureau of Reclamation, Denver, Colorado.

Peterka, A.J. (1984). *Hydraulic Design of Stilling Basins and Energy Dissipators*, Engineering Monograph No. 25, Eighth Printing, U.S. Department of the Interior, Bureau of Reclamation, Denver, Colorado.

Townsend, C.L. (1981). *Control of Cracking in Mass Concrete Structures*, Engineering Monograph No. 34, Revised Reprint, U.S. Department of the Interior, Bureau of Reclamation, Government Printing Office, Denver, Colorado.

Wisner, G.Y., and Wheeler, E.T. (1905). "Investigation of Stresses in High Masonry Dams of Short Spans," *Engineering News*, Vol. LIV, No. 6.

Knowledge and Water: Colorado State University's Water History

Neil S. Grigg[1]

Abstract

The roles of public educational institutions in western water development make an interesting chapter in histories of the Plains and Mountain West. Colorado State University's programs evolved from their frontier beginning to have recognized global impact. Early history of mostly–engineering programs is described, and the university's response to need for interdisciplinary programs to address today's issues is explained.

Public universities and water

Water issues are complex and generate conflict, as John F. Kennedy said: "Anyone who can solve the problem of water deserves not one Nobel Prize, but two—one for science and the other for peace." Another saying in the West, is "Whiskey is for drinking and water is for fighting." These attributes characterize the history of water development in the West, and Colorado State University's programs have evolved to respond to them, first to deal with the complexity, and more recently, to deal with conflict as well.

In a democracy, complexity and conflict in public issues cry out for education, transfer of knowledge, and forums to work out differences. Public universities serve as useful institutions with unique roles to help with public issues, and Colorado State's role is captured in the title of its history, "Democracy's College in the Centennial State" (Hansen, 1977). Land Grant Universities, such as Colorado State, with their focus on resource issues, are well–positioned to contribute to water solutions, and their contributions are recorded in many accounts. For Colorado State, water knowledge has been a central theme during its 125–year history, and this story is an interesting chapter in local and regional histories of the Plains and Mountain West.

---

[1]   Professor, Department of Civil Engineering, Colorado State University, Fort Collins, CO 80523; phone 9780-491-3369; neilg@engr.colostate.edu

In this short "water history" of Colorado State the goal is to present an account of the main participants and events in this rich period of western and university development. In writing the account, however, I am impressed with the difficulty of the historian's task in presenting objective accounts. The account emphasizes the role of engineers in water development, but in recent years other disciplines have made important contributions, especially to deal with more complex regulations, greater appreciation of ecology, and greater conflict between interest groups. These aspects are especially evident in 2002, because the region is facing its deepest drought in 50 years at the same time that demands on water have increased greatly. Because published accounts emphasize Colorado State's first 100 years, this history falls short in recognizing fields with more recent contributions such as atmospheric science, fisheries, earth resources, and the social sciences. Perhaps their histories can be written later.

How did Colorado State come to be involved with water?

Water science has evolved for thousands of years, and people may not think of it as a "high tech" field. However, when you combine weather, hydrology, water delivery and control, water quality, agronomy, and institutions that manage water, you end up with a complex mix. That was the case when settlers began farming in Colorado in the 1850s.

It is true that irrigation history began 6000 years ago, that early Rome was able to import water via its famous aqueducts, and that sophisticated water supply systems had already been built in the United States before Colorado State was established. However, farmers, miners, industrialists, and settlers in the West did not know all of the secrets of earlier civilizations or eastern engineers, and needed practical knowledge about water for western conditions.

Colorado State's work in water began soon after the university's founding in the 1870s. In 1883 Elwood Mead joined the faculty to teach mathematics. However, "...Colorado's urgent need for direction concerning the development and utilization of water resources and Mead's creative competence..." soon directed his work toward water and specifically toward water law and administration. Mead proposed and the College's Board accepted the creation of a course in irrigation engineering. This was the first course in water resources at Colorado State and may have been the first in the nation (Hansen, 1977).

Later, Mead, for whom the lake above Hoover dam is named, laid the foundation for Colorado State's hydraulic engineering program. He taught courses about water flow in canals, reservoirs and closed conduits. In 1888, he became Wyoming's first state engineer and wrote the first irrigation code for arid and semi-arid regions.

Irrigation and hydraulic engineering

Irrigation and hydraulic engineering soon became a field of study itself, and Ralph Parshall was one of Colorado State's first students in the new department of civil and irrigation engineering. He then became a Colorado State instructor in 1907. Later, he joined the U.S. Department of Agriculture and continued to work at Colorado State, where he and his colleagues drew recognition to Colorado State's contributions to irrigation technology and hydraulics laboratory work. In 1925 Parshall patented the Parshall Flume, which is used throughout the world.

Colorado State's President Charles A. Lory (1909 – 1940) had great influence on the school's water programs. Lory's family lived on a farm near Fort Collins, and by age 15 he was working on irrigation tasks and was soon to be employed as a ditch superintendent. Although interested in agriculture, he chose to become a teacher, and eventually returned to Colorado State Agricultural College to teach physics, mathematics, and electrical engineering. He made great contributions to the start of agricultural programs, including the Experiment Station and Extension Service. In 1912, he was instrumental in developing an agreement with the Department of Agriculture to locate the hydraulics lab at the college. This hydraulics lab later became a key instrument in research and education, helping train graduates who served all over the world.

During 1898-1905, investigations by the college's students had begun for the water development project now known as the Colorado–Big Thompson Project. In 1934 the Bureau of Reclamation, then headed by Elwood Mead, took over the project. The history of the CBT project is described in the "Last Water Hole in the West: the Colorado Big Thompson Project and the Northern Colorado Water Conservancy District," a text by Colorado State history professor Dan Tyler (1992).

During the 1930's President Lory's water leadership was recognized on the national scene. Harold Ickes, Roosevelt's Secretary of Interior, appointed him to head the Repayments Commission, which led the way to the Reclamation Projects Act of 1939. Later, he joined the National Resources Planning Board, and his name appears on river basin studies of the period.

During Lory's tenure, graduates from the college achieved distinction in the water industry. Many were leaders in the Bureau of Reclamation, and they participated in the federal government's most important water projects.

Post-war growth

After World War II, the stage was set for growth in Colorado State's water programs. Maurice L. Albertson arrived in 1947 to become a key player in program development. As president of his ASCE Student Chapter, Albertson had attended a national ASCE Convention in Denver in the summer of 1940 and reported that the conference was a major point in his life because of the papers and the USBR Hydraulics Laboratory that had moved from Fort Collins to the Custom House in downtown Denver. Albertson was fascinated by the hydraulic models and displays concerning water. Carl Rohwer and Ralph Parshall told Albertson that Nephi Christiansen, who was Dean of Engineering and Head of the Civil Engineering

Department, was planning a future graduate program in water resources at Colorado A and M (Albertson, 1997).

After his graduation from Iowa State in 1941, Albertson received a graduate research assistantship at the University of Iowa in Iowa City where he met Hunter Rouse. Paul Benedict of USGS and Emery W. Lane, later of the USBR, helped him. After a year at TVA, Hunter Rouse invited Albertson back to Iowa to work on wind tunnel research. Following graduation, Albertson joined the Colorado State faculty in 1947.

Colorado State's Hydraulics Laboratory has been a centerpiece of the water programs. Ralph Parshall and colleagues designed the first lab, which was completed in 1913. It led to many productive research projects by the university and government agencies located nearby. The Bureau of Reclamation used the lab to study plans for the Hoover, Grand Coulee and Imperial Dams.

Max Parshall, son of Ralph Parshall, had managed the Hydraulics Lab after the USBR left in 1938. The lab also provided a place for the Bureau of Public Roads to study drainage and flood control for the nation's interstate highways. In the early 1960s, the lab was moved from main campus to its present location at the Engineering Research Center, and it continues to provide unique experimental facilities for hydraulic research.

Initially, Albertson worked with Jack Cermak, his first M.S. student, and received funding from ONR for wind tunnel research. This led to establishment of the Fluid Dynamics and Diffusion Laboratory, a companion to the Hydraulics Laboratory. Later, Albertson was also instrumental in establishing a Hydromachinery Laboratory

Dean Peterson came in 1949 and helped greatly in fund–raising and faculty recruitment. When India and Pakistan gained independence there was a great push to develop water project models, and several key dams were tested in the Hydraulics Laboratory. Al Peterka, a staff member at USBR, was a key advisor in lab work. President William Morgan (1949–1969) let the lab keep overhead funds to build the program. In its early years, federal agencies used the lab for research. Paul Benedict of USGS, Carl Izzard of the BPR, and Whit Borland of the Bureau of Reclamation were key collaborators.

President Morgan presided over the period when Colorado State's water programs experienced their rapid post–war growth and were expanded globally. He served as Chair of the Water Committee of the National Association of State Universities and Land Grant Colleges at the time of the passage of the Water Resources Research Act. He was followed by A. R. Chamberlain (1969-79), who with a specialty in water within the Department of Civil Engineering had been Colorado State's first PhD graduate in 1955. Chamberlain also served on the NASULGC water committee and provided steady support to water programs. During the tenure of Colorado State's current President, Albert C. Yates, a Colorado Commission on Higher Education award for excellence in water resources programs was secured and the university's

Program of Research and Scholarly Excellence designation was conferred to water programs. Also, the university's "Water Plaza" was built and "Water Archives" initiated in the Morgan Library during his tenure.

Hunter Rouse, former Dean at the University of Iowa and the engineer for whom ASCE's Hunter Rouse Hydraulic Engineering Lecture is named, gave his first summer course at Colorado State in 1940. Rouse observed the rapid growth of Colorado State's water programs and decided that the factors behind the growth should be documented before the information was lost (Rouse, 1981). Rouse continued to teach his summer course and inspired several generations of faculty and students.

After joining the USGS staff in Fort Collins in 1957, Daryl B. Simons became a Professor and Associate Dean at Colorado State. He contributed to and established a well–known program in sedimentation and river engineering. The hydraulics team at the Engineering Research Center, including other professors such as Everett Richardson and Hsieh-Wen Shen, completed many model studies and research investigations.

Vujica Yevjevich joined Colorado State in 1957, after a research career in his native Yugoslavia. He established a well–known program in hydrology and stochastic hydrology, and advised a number of graduate students who went on to distinguished careers in the US and other countries. In 2002, one of his students, Ignacio Rodriquez–Iturbe, won the Stockholm Water Prize. In that same year, another student, Francisco Gomide, was named Brazil's Minister of Mines and Energy. Another Yevjevich student from Brazil, Jerson Kelman, had earlier been named Director of Brazil's national water agency.

Colorado State's groundwater faculty also made important contributions. Art Corey established a Groundwater Laboratory at the Engineering Research Center. Another key program that spun out of the engineering water program was the university's Atmospheric Sciences Department, a national leader among atmospheric science departments.

Although this short account does not do justice to non–engineering work, it would be remiss not to mention the fertile climate for interdisciplinary work that emerged on the campus after the 1960s. This enabled Colorado State to attract scientists who had already achieved highly in public sector work. For example, Henry C. Caulfield, former Director of the US Water Resources Council, served as a Professor of Political Science and interacted regularly with the engineers and natural scientists.

International work

Colorado State's international water programs began in the mid-1950s, when the university helped develop graduate-level water programs at the University of Peshawar in Pakistan, where the university's influence is still felt more than 30 years later, especially for irrigation systems and on-farm water management. Similar projects were conducted in Afghanistan. In 1959, Colorado State helped establish the Southeast Asia Treaty Organization graduate school in Thailand. The school is now known as the Asian Institute of Technology. In the early 1960's Colorado State researchers, under Maurice Albertson's leadership, had an active part in creating the Peace Corps.

Colorado State has been active in international training and in 1967 established the International School for Water Resources. In the 1970s, Victor Koelzer retired from his post as Chief Engineer of the National Water Commission and assumed the directorship. Koelzer introduced a course on water resources planning, modeled after his experience with USBR and Harza Engineering Company. Later, Warren A. Hall, a pioneer in water resources systems engineering and former Director of the Office of Water Resources Research, joined the faculty and took over the school. In its first 25 years, the ISWR organized training programs for over 350 professionals from 57 nations.

Colorado State faculty assisted with several irrigation management projects in Egypt's Nile Valley. Everett Richardson and Dan Sunada directed key projects in the series. A $25 million contract awarded by the Egyptian government was the culmination of over two decades of work. About 40 Colorado State faculty members provided technical assistance to the project. About 55 Egyptian students enrolled in graduate programs in the United States; 30 at Colorado State University. Colorado State was also active in Pakistan irrigation research, and operated projects for the Agency for International Development. One of these, the Water Management Synthesis Project, was directed by Wayne Clyma and produced many reports about irrigation technology and management.

The Colorado Water Resources Research Institute

After over 125 years of work in the field of water, Colorado State's programs have diversified to the point where 25 departments participate in multi–disciplinary activity, much of which is coordinated by units such as the Water Center and the Colorado Water Resources Research Institute. CWRRI was one of the state institutes established by Congress in 1964. The Colorado Legislature authorized it in 1981 and again in 1997. It began as the Natural Resources Institute with Stephen Smith as the first director. Later, Norman A. Evans was director until 1988. He was followed by Neil S. Grigg and then Robert C. Ward. Colorado State created the Water Center to take on integrating functions within the university, and now CWRRI and the Water Center are combined. CWRRI has published over 300 research reports, and is a valuable entity to the state's water managers. Colorado State's annual partnership with water utilities and districts has sponsored the Children's Water Festival and several other outreach events. One of these, Hydrology Days, was started by Hubert Morel–Seytoux in the 1980s and continues under Water Center sponsorship.

Toward the future

Today, water activity at Colorado State is even more extensive than the past. The university has over 100 courses related to water, and over 900 scientists work in government natural resource research agencies in Fort Collins. The university's water programs address more complex issues than in earlier times. Roles of public universities have blended into a more complex society with more and different institutions plying their knowledge to solve problems of water. Colorado State's water programs remain vibrant and continue to evolve in response to the demands of a changing world.

References

Albertson, Maurice L. 1997. My 50 years at Colorado State University. Department of Civil Engineering. Colorado State University, Fort Collins, CO.

Hansen, James E. II. 1977. Democracy's College in the Centennial State: A History of Colorado State University, Fort Collins, CO.

Rouse, Hunter. 1981. Hydraulics, Fluid Mechanics, and Hydrology and Colorado State University. Engineering Research Center, Colorado State University, Fort Collins, CO.

Tyler, Daniel. 1992. Last Water Hole in the West: the Colorado Big Thompson Project and the Northern Colorado Water Conservancy District, University Press of Colorado, Niwot, CO.

The Iowa Institute of Hydraulic Research[1]: Seventy-Plus and Still Going Strong

Robert Ettema[2], Cornelia F. Mutel[3], and V. C. Patel[4]

Abstract

This paper presents a overview of the Iowa Institute of Hydraulic Research (IIHR). The overview outlines IIHR's beginning as a single-room hydraulics laboratory in 1920, charts its growth, and describes its current research and educational prospects. Periods of development are discussed briefly in terms of IIHR's sequence of directors. Over time, the range of IIHR's research and educational activities has widened. It is intriguing, though, to observe that many of the opportunities IIHR currently faces are intrinsically the same opportunities that led to IIHR's founding. In particular, IIHR's research and educational activities continue to be motivated by the overall need to understand and simulate water flow, as well as to design diverse structures that influence water flow. A recent book, *Flowing Through Time*, provides detailed coverage of IIHR's founding, its development, and its present prospects.

Introduction

The Hydraulics Laboratory, administrative center for the Iowa Institute of Hydraulic Research (IIHR), rises as an imposing red-brick edifice on one of the busiest corners of the Midwestern college town of Iowa City, Iowa (Fig. 1). Located alongside the Iowa River near the downtown area on The University of Iowa's central campus, this building is seen by thousands every day. Only the initiated, those who are hydraulicians, hydrologists, or fluid mechanicians with some inside association, are likely to realize that the building's walls, if they could talk, would tell the ageless story of people attempting to understand and control that most common of the earth's constituents, water.

Part of that story is told in a recent book on IIHR's history, *Flowing Threw Time* (Mutel, 1998). IIHR, like other hydraulics laboratories, has performed research crucial to the workings of modern society. That research dated from early in the century to present times. As the activities and personalities that have energized this

---

[1] Now renamed "IIHR – Hydroscience & Engineering"
[2] Research Engineer and Professor, IIHR, The University of Iowa, Iowa City, IA 52242.
[3] Historian with IIHR, The University of Iowa, Iowa City, IA 52242.
[4] Director and Professor, IIHR, The University of Iowa, Iowa City, IA 52242.

Figure 1. IIHR's Hydraulics Laboratory following its 2001 renovation.

organization since 1920 took hold, IIHR grew into of one of the U.S.'s major hydraulics laboratories.

This paper sketches out IIHR's creation, development, and present prospects for advancement. A common feature throughout the outline is IIHR's leadership by directors known internationally for their contributions to hydraulics research and education. The birth of IIHR is largely attributable to one man, Director Floyd Nagler. IIHR was his brainchild. IIHR's growth to maturity was guided and motivated by Directors Hunter Rouse and John F. Kennedy, both of who were internationally prominent hydraulicians. Its current development is presided over by Director V.C. Patel, an internationally known fluid mechanician. Since it is often asked how IIHR began and what IIHR's future holds, this article focuses on the first and current directors (Nagler and Patel). Detailed information about Directors Rouse and Kennedy is presented elsewhere (e.g. in the book *Flowing Through Time*).

Another common feature of IIHR's research is its inclusion of the diverse facets of water resources. From its creation, IIHR undertook research into water flow in rivers, water movement through the hydrologic cycle, the design and performance of various hydraulic structures, and the fundamentals of fluid flow. The breadth of IIHR's interests in water and fluid flow recently led IIHR to adopt a new name – IIHR–Hydroscience & Engineering – to reflect the scope of its research and educational activities.

Birth of an Institute

IIHR's origins lie in the foresight of the faculty of The University of Iowa's Department of Mechanics and Hydraulics who influenced the design of the Burlington Street dam on the Iowa River. They enabled the construction in 1919 of a small hydraulics laboratory with a 40-m-long and 3-m-wide open-channel flume. The small, one-room laboratory expanded very rapidly, and the basic structure of the current Hydraulics Laboratory was completed in 1932. It featured, in addition to the previously mentioned large-scale open-channel flume, a large constant-head tank in the top story and a recirculating water system permitting small-scale model and laboratory testing on all floors of the building. The scope and complexity of the problems undertaken by the laboratory soon received national attention. They prompted the formal organization of IIHR in 1931 as "an agency for the coordination of talent, facilities and resources available through The University of Iowa to be brought to bear on problems in hydrology and hydraulic engineering."

It is said that IIHR's first director, Floyd Nagler, was a man of boundless energy and drive, capable of balancing ten activities simultaneously and keeping them all going. Nagler's primary legacy before his premature death at age 41 was the establishment of the physical and organizational structure of the Iowa Institute of Hydraulic Research. He had been brought to Iowa to foster the research of a tiny laboratory with under 500 square feet of floor space; he left a five-storied laboratory well over 50 times as large. He guided the transition of this Hydraulics Laboratory to the full-blown IIHR and served as the founding director of both the laboratory and the institute. As the first to steer the vessel, he set IIHR on its course and established the traditions that were to become its hallmark: traditions of high productivity, excellence, innovation in research, and engineering service.

His drive led to an amazing proliferation of research that exceeded even his expectations. At first, research work at the Hydraulics Laboratory was synonymous with work completed by Nagler: he was, after all, the only researcher present. As his staff grew, Nagler remained involved in the Hydraulics Laboratory's research as well as its administration, and he seemed to have a thorough grasp of IIHR's diverse efforts and their import. His own efforts in later years focused on surveys regarding the water-resource potential and flow characteristics of the Mississippi River and its watershed. This research was performed for the Army Engineers (now the U.S. Army Corps of Engineers) and for Iowa's Fish and Game Commission and Board of Conservation.

Early Development, 1920–1933

During Nagler's tenure, the field of hydraulics rapidly expanded in importance and breadth of activity. Hydraulics-related problems were abundant and obvious. As Nagler himself wrote, "The field open to hydraulic research is almost unlimited." Laboratories were popping up across the country at major universities, encouraging the assemblage of collaborating researchers, and enabling the evolving use of small-scale models, which in turn stimulated an attack on basic problems that had not previously been investigated. Spin-off projects were attacked with zest, for hydraulic research was providing useful and necessary answers to basic questions and thus was well funded. All these factors combined to feed one another, the funding permitting the expansion of laboratories and personnel, which jointly fostered the more sophisticated use of models, resulting in a recognition of additional research problems that beckoned the researchers ever onward. Even the most straightforward problems commanded a sense of excitement and energy.

Research during this period focused on questions that until then had remained largely unaddressed. During Nagler's tenure, IIHR attacked all these questions, typically under contracts from a government agency, private company, or engineering firm that desired their applied results (although IIHR graduate students remained engaged in more theoretical challenges or, as Nagler explained, "in the study of more academic problems"). Nagler and his colleagues examined the water-carrying capacity of various culvert shapes and materials. They tested the performance of meters that measured current, assessed characteristics of the hydraulic jump, and investigated how metal flap gates affect water flow. They observed how bridge piers and abutments of various shapes block or constrict water, causing local scouring of the riverbed. Sometimes the most basic questions revealed surprising results; for example, Nagler's team discovered that in a given period, almost twice as much water could pass through a smooth concrete culvert as through a corrugated metal culvert with the same inner diameter. Or (as Nagler pronounced with pride), the 57 unlikely solutions that had been proposed to abate flood problems at Milan, Illinois, had been reduced at the Hydraulics Laboratory into a single successful resolution.

Model testing for the Army Engineers' Mississippi River locks and dams would become one of the Hydraulics Laboratory's major focuses in the 1930s, bringing in a substantial portion of its consulting revenue. Thus, Floyd Nagler and IIHR, in its early years, were instrumental in the massive restructuring of the Upper Mississippi that today characterizes the river and its use.

Attaining Stature, 1940s – 1980s

Following the death of Professor Nagler in 1933, Director and Associate Director positions, respectively, were filled by Dean C.C. Williams and Professor F. T. Mavis until 1936, then by Dean F. M. Dawson and Professor E.W. Lane until 1942. During these two periods, graduate student involvement in research increased significantly. Also, IIHR initiated the sponsorship of a national hydraulics conference, setting a tradition which continued over twenty years with seven conferences, and ultimately leading to the Hydraulics Specialty Conference series, now run as the Annual Water-Resources Engineering Conference by the American Society of Civil Engineers (ASCE).

Professor H. Rouse became Associate Director in 1942, then Director in 1944. He reorganized IIHR to consist of an active research staff, drawn from Engineering College faculty. He redirected IIHR toward basic research and graduate study, and he wrote several textbooks, some with the cooperation of IIHR's staff. These activities coincided with the emergence of hydraulics and hydrology as separate scientific disciplines through adaptation of advances being made in fluid mechanics in general, and hydrodynamics and aerodynamics in particular, and with the expansion of research funding by government agencies brought on by World War II. In 1947, IIHR again expanded with the addition of a laboratory annex and wind tunnel (IIHR's former West Annex). Rouse is credited widely with considerably enhancing IIHR's national and international prestige as a premier fluids-engineering laboratory.

Professor J.F. Kennedy became director in 1966. In response to the challenge of operating IIHR during a period when research funds were becoming considerably more difficult to obtain (primarily because of the formation of other laboratories, often staffed by IIHR graduates), he broadened IIHR's research scope to maintain a fertile balance of basic and applied research. As a result of Kennedy's leadership and his own technical reputation, IIHR thrived and expanded to its current level of activities and facilities. Upon Kennedy's resignation as director in July 1991, Professor R. Ettema was appointed Acting Director.

IIHR's Research and Educational Challenges in the Years 2000-Plus

Professor V.C. Patel was appointed as IIHR's director in June, 1994, at a time when the scope of IIHR's activities was broadening in response to evolving societal interests, new technologies and new insights, and coincident interests and expertise among the people comprising IIHR. That broadening of scope drew IIHR further into advanced aspects of hydrology and hydrometeorology, computational fluid dynamics, biofluid mechanics, and environmental concerns, and also kept IIHR busy with hydraulics research. In response to these and other developments, Patel encouraged both the renaming of IIHR as "IIHR - Hydroscience & Engineering" and further broadened its research and educational focus. In 2002, together with Associate Director Tatsuaki Nakato, Patel established an IIHR field laboratory on the Mississippi River near Muscatine, Iowa. Through Patel's leadership, IIHR has been readied to further address evolving and new societal needs and basic hydraulics questions, as discussed below.

Evolving Societal Interests. IIHR has long been attuned to changing societal interests and needs, and from them has created new possibilities for research and education. Doing so has proved crucial to society because the relevance and importance of hydraulics – that is the practical and pervasive role of water in human affairs – has never diminished. Over the years, IIHR has provided national and worldwide leadership in areas such as thermal pollution in water bodies, power-plant cooling, river control structures, ice engineering, and viscous ship hydrodynamics.

Presently, for instance, hydraulics has progressively turned "green" in response to concerns about environmental well-being. Traditionally, hydraulics has been associated with manipulation of natural water bodies (rivers, aquifers, and coasts in particular) to meet societal demands. Though these activities remain critically important, society perceives them a little skeptically -- especially those societies whose water needs have by-and-large been met. In response to public perception and to real needs, contemporary hydraulics is placing heightened emphasis on understanding and enhancing the natural environment. Research and teaching in the areas of environmental hydraulics (or environmental fluid mechanics), habitat hydraulics, and ecological hydraulics have become topical in response to concerns about environmental well-being.

Increased interests in efficient water use compel research and education in water-resources engineering. Population growth, together with concerns for environmental well-being, are placing increasing demands on water supply. The term "water stress" signifies the limiting influence of water availability on the expansion of human communities, especially mega-cities like New York, Los Angeles, Tokyo, and Cairo. Water's abundance or scarcity is even felt in much smaller communities like Iowa City, which faces difficult decisions concerning the feasibility of expansion. For hydraulics, water stress implies a host of research prospects including new types of conveyance and hydraulic structures (particularly for large cities and arid areas); improved flow-control and flow-regulation systems; greater efficiencies in power-production systems using water (as most do); water recycling; and innovative integration of natural water systems (e.g. wetlands) into human wastewater systems.

New Technologies, New Insights.   Technological innovations facilitating new insights into fundamental aspects of fluid flow are creating opportunities for hydraulics research and practice.   At IIHR, hydraulics is becoming increasingly sophisticated and rigorous, progressing substantially beyond the hydrodynamicist Theodore von Karman's somewhat derisive characterization as "the science of variable coefficients."   Substantial improvements in the design and control of numerous hydraulic and hydrologic systems undergirding social infrastructure (e.g. those for handling storm-sewer flows, hydropower generation, or irrigation) are in the wind or already underway. Hydraulics is abuzz with "hydroinformatics." Advances in computational hydraulics and computational fluid-dynamics (the two are merging), augmented by instrument development, are facilitating those improvements. Ultimately, they will supercede present-day, often empirical and seat-of-the-pants approaches used to design and operate hydraulic systems.

Several current projects at IIHR illustrate the opportunities presented by technological developments and by advances in the understanding and modeling of fluid flow.   One project involves application of multi-component laser-Doppler velocimetry (LDV) to obtain detailed information on flow within hydropower intakes and turbines. The project is motivated by two concerns: fish passage through hydropower plants, and the occasional need to re-aerate oxygen-deficient flows released from reservoirs.   This project also involves application of advanced computational fluid dynamics (CFD) modeling of the sort originally developed by fluid dynamicists for simulating and investigating flow around airfoils and ship hulls. In a similar vein, LDV instrumentation and CFD simulation are being used at IIHR to investigate flow in dune-bed channels (albeit still with two-dimensional dunes) and flow in water-intake pump sumps.   Another promising tool, particle-image velocimetry (PIV), is beginning to make possible, in virtually the twinkling of an eye, flow imaging and measurement that until now have been extremely arduous. In the long run, application of powerful CFD methods and convenient instrumentation will completely change the way nonstandard hydraulic structures are designed. Those innovations already are occurring in the design of ships, planes, and hydromachines. The change will entail collaboration among engineering disciplines.

Coincident Interests and Expertise.   An issue of growing significance is the importance of professional interaction among disciplines with interests and expertise that coincide. Such interaction, applied to larger problems and processes, is not easily facilitated with today's tendency toward increased specialization. It is an issue, therefore, with implications concerning the ways hydraulics and related topics are taught. In this respect, a strength of IIHR has been its capacity to blend personnel, facilities, and activities that elsewhere are found in diverse departments with little or no interaction or shared goals among them. A faculty and research staff drawn from diverse backgrounds, dedicated to basic research and the solution of real-world problems, is the hallmark of IIHR. Joint projects and innovative approaches to research are readily conceived and executed in this environment.

Present Prospects

IIHR's research expertise lies in core fluids-engineering disciplines that will be relied upon to solve problems of the air and water environment. With renewed national emphasis on human wellness and related research in health, biofluid dynamics offers vast research opportunities for IIHR. IIHR has been involved in this area only on a limited basis in the past.

A growing number of emerging projects engaging IIHR are multidisciplinary in nature. Because water and fluid dynamics play major roles in diverse aspects of human affairs and the natural environment, IIHR will have to become increasingly engaged in research and development efforts that are primarily multi-faceted and multidisciplinary.

IIHR has long enjoyed a worldwide reputation for its educational and research programs. This is a fortunate circumstance as problems related to water and the environment are receiving increased attention around the globe, and IIHR is in a unique position to provide leadership at the international level. The problems are particularly acute in the so-called Rapidly Developing Countries where rampant industrialization has strained water resources and quality, and also in the Third-World Countries aspiring to increase industrial production in fragile environments with limited water resources. Moreover, these problems are not confined by international boundaries. With its extensive international connections, its expertise in hydraulic engineering and water resources, and its growing capabilities in mathematical modeling of complex systems, IIHR is uniquely qualified to develop an international program of research and education in these areas. IIHR has become a world leader in the development of mathematical models of river networks and associated hydraulic structures. These models incorporate state-of-the-art computing and data-processing methods as well as modern developments in hydrology, hydrometeorology, and remote sensing with ground-, air-, and space-based instrumentation, resulting in tools that are needed to attack large-scale international water-related problems.

Concluding Remarks

Ongoing changes in national and international priorities, developments in technology and knowledge, and the coincidence of people's interests have enabled IIHR to configure effectively its research and educational programs over time. Presently, IIHR's prospects reflect increased emphasis on multidisciplinary research in fluid dynamics, water-resources engineering, and the hydrosciences. Additionally, IIHR intends to strengthen its partnerships with industry. These actions position IIHR to continue to contribute in a major way to the solution of water-resource problems around the world.

References

Mutel, C.F. (1998). *Flowing Through Time: A History of the Iowa Institute of Hydraulic Research,* Iowa Institute of Hydraulic Research, The University of Iowa, Iowa City, IA. 299 pp. (This book is available for purchase through IIHR's web site: <http://www.iihr.uiowa.edu/products/history/index.html>).

Nakato, T., and Ettema, R. (1996). *Issues and Directions in Hydraulics,* A.A. Balkema, Rotterdam Netherlands, Brookfield, Vermont. 495 pp.

Evolution of NRCS Hydrologic Methods through Computers

Donald E Woodward, Member ASCE[1]

Abstract

The Natural Resources Conservation Service (NRCS), formerly known as the Soil Conservation Service (SCS), has developed hydrologic procedures for ungaged watersheds. NRCS initially used manual procedures to design conservation measures at the farm level. The agency was authorized to make small watershed studies in 1954 and this encouraged NRCS to develop manual procedures that could be used for ungaged watersheds across the United States. As computers became available, NRCS used them to develop and refine hydrologic procedures. This paper will explain the evolution of the current hydrologic procedures.

History

The Natural Resources Conservation Service (NRCS) was established in 1933 as the Soil Erosion Service with the primary purpose of reducing soil erosion (Rallison 1980) and became the Soil Conservation Service (SCS) with the passage of the Soil Conservation Act of 1935. SCS became NRCS in 1994.

In 1933, NRCS began to develop manual procedures to design measures to provide soil and water conservation at the farm level. These procedures included methods to provide peak flow estimates and tended to vary by regions. Designers used what was available in the literature.

The Flood Control Act of 1936 required NRCS to develop hydrologic procedures at the watershed level. It authorized NRCS to make improvements in 11 watersheds. NRCS worked with various consultants and experts within the agency to develop procedures that provided methodologies for analysis of alternatives and design of conservation measures. NRCS concentrated on development of infiltration losses curves that were typical of the watershed being investigated. NRCS also used phi

---

[1] National Hydraulic Engineer (Retired), Natural Resources Conservation Service, 7718 Keyport Terrace, Derwood, MD 20855: Phone 301-977-6834: dew7718@erols.com

index procedures. These procedures never gained universal popularity or use. (Woodward 2002)

In the 1940s, NRCS felt that it could use L.K. Sherman's concept of plotting direct runoff versus rainfall to estimate runoff. NRCS built on this idea by proposing to use a regression equation that contained the parameters of soil type, land use, antecedent rainfall, storm depth, average annual temperature, and date of storm. This would have been a co-axial graph and would have been limited to gaged watersheds that were similar to the ungaged watershed being studied. The regression equation never gained wide use and required stream gage information (Woodward 2002).

The passage of the Watershed and Flood Preventions Act (PL-556) authorized NRCS to make watershed studies. NRCS decided it needed a simple, standard hydrologic procedure that was applicable across the United States. NRCS began to build on ideas that were in the literature. Most of the procedures that were available were designed to be used in a manual mode. NRCS developed the rainfall-runoff equation based on rainfall-runoff data from 24 small watersheds. These watersheds were located across the United States and were primarily agricultural. (Rallison 1980 and Woodward 2002)

It was decided that to determine the impacts of various alternatives in the watersheds, current manual procedures should be used. These manual procedures included storage indication method of reservoir routing, Wilson channel routing procedures, and the unit hydrograph concept. These procedures with the NRCS runoff equations allowed the practicing hydraulic engineer, together with the economist and geologist, to evaluate the alternatives within an ungaged watershed. (Woodward 1973)

TR-20

In 1965, NRCS developed a computer program to evaluate the effects of proposed structural measures on peak flow. This project evaluation model was developed under contract with a private software supplier. The contract with C.E.I.R. was developed using standard NRCS hydrologic techniques and procedures (C.E.I.R 1964). The program was written to operate on the IBM 360 or a mainframe computer. The contract also specified that the current NRCS procedures would be the driving force rather that the computer program. Thus, the computer program operated in the same manner as the practicing hydraulic engineer. I surmise that this helped make the program very popular with hydraulic engineers. It was felt that the resulting computer program would provide the hydraulic engineer with a tool that would allow more alternatives to be evaluated and to reduce mathematical errors, but not reduce that time needed to make a watershed study. The resulting computer program was known as "Project Formulation Program – Hydrology," Technical Release 20 (TR-20). (SCS 1965)

The initial TR-20 computer program used the control word concept to organize the input data and determine which subroutine the computer would use. These control words were known to the hydraulic engineer. Sample control words include:

RUNOFF	develop a hydrograph
RESVOR	route a hydrograph through a reservoir
REACH	route a hydrograph through a channel reach
ADDHYD	addition of hydrograph
SAVMOV	save the hydrograph and move to a storage location
DIVERT	divert the flow out of the watershed or to other location

This feature was intended to make the computer program easier to use and consistent with normal manual procedures.

The initial NRCS unit hydrograph had a triangular shape (SCS 1986). The triangular shape allowed for the development of the runoff equation (1):

$$q_p = \frac{0.208QA}{D/2 + 0.6Tc} \tag{1}$$

Where q_p is the peak rate of runoff in m^3/sec
      Q is the volume of runoff in mm
      A is the drainage area in sq km^2
      Tc is the time of concentration in hrs.
      D is the duration of the storm increment in hrs.
      0.208 is the conversion and shape factor depending on the volume in the rising limb of the hydrograph

Also, it was easy to accumulate triangular hydrographs when manually determining the storm hydrograph. Normally, triangular hydrographs were drawn on graph paper and then the individual hydrographs were accumulated. The advent of the computer allowed NRCS to use the curvilinear unit hydrograph shape and to accumulate the individual unit hydrographs to form the storm hydrograph in a systematic manner. NRCS hydrologic procedures in TR-20 now use unit hydrographs that can be varied by region or storm. Normally, 24-hour storms were used in the manual procedures. The TR-20 computer program allows the hydraulic engineer to use one of the four 24-hour rainfall distributions, a historical rainfall distribution, and a specific design distribution.

The storage indication method of reservoir routing is used in TR-20. Normally, the working curves were developed manually and the actual routing was done manually. The computer program was designed to develop working curves for many alternatives and to check the curves to prevent errors in the routings. This provided the user with more options to study.

It was soon realized that the channel routing procedure being used was not conducive to computer logic. NRCS developed the convex method of channel routing. This method provided a logical and consistent method for moving the hydrograph downstream. While the convex method provided answers, it was soon realized that the procedure tended to over route in long reaches and under route in short reaches. It was decided to use another reach routing routine, Mod Att-Kin. This reach routing routine improved the shortcomings of the convex method. (SCS 1986)

NRCS decided to use the Muskingum-Cunge method of reach routing. The computer has allowed the user to take advantage of the trial and error requirements of this reach routing routine.

Before the development of the computer program TR-20, the hydraulic engineer normally evaluated only two or three combinations of structural measures. He/she normally used a minimum of stream cross sections to determine the economic benefits of the proposed structural measures. The computer program allows the hydraulic engineer to study all possible combinations of structural measures with all the available cross sections. The current version of the TR-20 computer program develops hydrographs for up to nine different storms and performs routing though 120 reaches and 60 structures. This is way beyond the normal combinations of alternatives that a hydraulic engineer would investigate manually. The summary of the various alternatives is presented in an easy to read format.

Soon, NRCS will release a windows version of the TR-20 computer program known as WinTR-20. WinTR-20 includes an easy to use input program that allows for checking of the data and allows the user to input the data in any sequence. The hydrograph development routine has been modified to simplify the input of the data. The user can design the output report and include graphical display of the output.

TR-55

In the 1980s, NRCS was asked to develop a procedure that showed the impact of urbanization on peak flows and the impacts of proposed stormwater management measures. It was decided to develop a manual procedure that quickly would provide peak flow estimates and short reservoir routing estimates using current NRCS procedures. (SCS 1986)

The TR-20 computer program was used to develop the peak flow curves in "Urban Hydrology for Small Watersheds," Technical Release 55 (TR-55). These curves are graphs of Tc versus csm/inch of runoff for the four standard NRCS 24-hour rainfall distributions and are in Chapter 4 of TR-55. The TR-20 computer program also was used to develop the approximate reservoir routing graph in Chapter 6 of TR-55. This graph provided a simple approximate shortcut procedure for reservoir routing for the four standard rainfall distributions. The tables for the Tabular Hydrograph Method in Chapter 5 of TR-55 also were developed from routings of the TR-20 computer program.

The TR-20 computer program is the backbone of TR-55, of which over 300,000 copies are in circulation. Shortly after the issuance of TR-55, NRCS began the development of the TR-55 computer program.

The TR-55 computer program uses the manual procedures in TR-55. The TR-55 computer program, while having procedures to determine the peak flow in watersheds, has several important limitations from TR-20.

> Tc is limited to 0.1 to 10 hours using the graphical peak discharge method
> Tt is limited to 3 hours using the tabular tables' method
> Tc is limited to 2 hours using the tabular tables' method

The procedures in TR-55, while basically manual in nature, have been automated for the PC environment. These manual procedures were developed by the TR-20 computer program. NRCS soon will release a windows version of TR-55, known as WinTR-55. This version will use the same basic limitation as the present version of TR-55 and will use the TR-20 computer program as the engine to determine the peak discharge.

The WinTR-55 computer program will have the capabilities of graph output. The WinTR-55 computer program uses the velocity method to compute Tc. It also has the capabilities to compute limited channel hydraulics. The reservoir storage and outflow rating curves are computed within the computer program. WinTR-55 includes an easy to use input program that allows for checking of the data.

Chapter 2 EFM

The hydrologic procedures in TR-55 were modified to be used as a tool for the design of conservation measures at the farm level. The Chapter 2 EFM (Engineering Field Manual) uses the graphical peak discharge procedures in TR-55 to estimate peak flow. Chapter EFM 2 is limited to a drainage area of 2000 acres and the method of estimating TC is the lag procedure developed by NRCS. This computer program is the backbone of the tools used by NRCS field offices. Chapter 2 EFM computer program computes the Tc for the watersheds given the drainage area, slope of the watershed curve number, and the length of the main watercourse. The other NRCS computer programs require the user to compute the Tc outside of the program. (SCS 1988)

SITES

As part of watershed analyses, NRCS developed a computer program to facilitate the design of floodwater retarding structures. This program, initially known as DAMS2, incorporated normal hydrologic procedures used to proportion earth dams (SCS 1971). Initially, NRCS used manual procedures to develop a 10-day hydrograph to determine the required floodwater storage and to develop a 6-hour hydrograph to proportion the emergency spillway. The DAMS2 computer program used computer techniques to develop and to route these hydrographs through the earth dam. DAMS2 computer program initially included a single watershed above the earth dam sites being investigated. It was revised to include the option to subdivide the upstream watershed. This option was only possible because of the computer.

In the 1990s, NRCS decided to upgrade the DAMS2 computer program to include new hydrologic techniques and to include head cutting routines to estimate if the earth spillway would breach. The computer program again allowed for incorporation of trial and error procedures into a computer program that normally would not be done manually. The SITES computer program now incorporates the same hydrograph development routines as in TR-20.

New versions of NRCS computer programs are available on the Web at www.wcc.nrcs.usda.gov/water/quality/hydro/Tools___Models/tools___models.html

Conclusion

NRCS developed manual hydrologic procedures for the design and evaluation of conservation measures. The computer has allowed NRCS to refine and improve these procedures into computer programs that are recognized worldwide for their technical excellent.

References

C.E.I.R., (1964). Computer Program for Project Formulation – Hydrology, C.E.I. R. Inc., Arlington VA

Rallison, Robert E., (1980). Origin and Evolution of the SCS Runoff Equation, Proceedings of the Symposium on Watershed Management '80, American Society of Civil Engineers, Boise ID

Soil Conservation Service, (1986). "Hydrology," National Engineering Handbook, Section 4, Washington DC

Soil Conservation Service, (1965). Project Formulation Program – Hydrology, Technical Release, Washington DC

Soil Conservation Service, (1971). Computer Program for Project Formulation – Structure Site Analysis, Washington DC

Soil Conservation Service, (1988). Estimating Runoff and Peak Discharge, Chapter 2 Engineering Field Manual, Washington DC

Woodward, Donald E., (1973). Hydrologic and Watershed Modeling for Watershed Planning, Transactions of American Society of Agricultural Engineers, Vol. 16 No 3, pp 582-4, St. Joseph MI

Woodward, Donald E., (2002). Curve Number Methods: Origins, Application and Limitations, Proceedings of the 2nd Federal Interagency Hydrology Modeling Conference, Las Vegas NV

HISTORY OF THE ILLINOIS STATE WATER SURVEY

Derek Winstanley[1], Nani G. Bhowmik[2], Stanley A. Changnon[3], and Mark E. Peden[4]

Abstract

The Illinois State Water Survey has provided leadership in establishing a scientific and engineering basis for water resources planning and management in Illinois and the nation since 1895. Although primary program areas have changed over time, some core programs remain. Program areas are monitoring, research, and public service, using the hydrologic cycle as the unifying framework. Major strengths of the Survey have been, and continue to be, partnerships with the other Scientific Surveys, the governing Board of Natural Resources and Conservation, the University of Illinois at Urbana-Champaign, Southern Illinois University, state and local government agencies, private companies, and professional organizations, especially the American Society of Civil Engineers. The Water Survey's contributions in the field of Civil Engineering are many and varied. These include river hydraulics, sedimentation and sediment transport, navigation impacts, hydraulic geometries of rivers and streams, urban runoff programs, groundwater evaluation analyses and modeling, wetland hydraulics, and water quality evaluations of streams, rivers and lakes. A goal is to make all of the Survey's major historical and current data sets and reports available via the Internet.

Introduction

Illinois is perhaps unique in the nation in having four Scientific Surveys to provide data and information and research results in the areas of water and atmospheric resources, natural history, geology, and waste management. In addition to meeting state needs, the Scientific Surveys participate in many regional, national, and international programs, and provide scientific and engineering data and information to a broad audience of researchers, professional organizations, governments, private companies, and individuals.

The Scientific Surveys can be traced back to the motivation and far-sightedness of

[1]Chief, [2]Principal Scientist Emeritus, [3] Chief Emeritus, [4] Director of External Relations and Quality Assurance, Illinois State Water Survey, 2204 Griffith Drive, Champaign, IL 61820-7495; phone 217/244-5459; fax 217/333-4983; dwinstan@uiuc.edu

19th century pioneers of the Midwest. Eminent scientists were successful in convincing public officials of the benefits of investing state funds in scientific institutions (Hays, 1980). The task of the Scientific Survey Chiefs and their staffs was "to require studies looking toward the future and not solely to solving the pressing needs of the present. Such long-term planning is quite unique among government agencies where the demands of the next election too often cause a concentration on short-term crises. The determination that these surveys should be devoted to study and analyses rather than regulatory administration is partly responsible for their uniqueness among state agencies" (William L. Everitt, in Hays, 1980, p. vii-viii).

When Illinois became a state in 1818, only 40,258 people lived in the state (Hays, 1980). By the end of the 20th century, the population had grown to more than 12 million and Chicago had mushroomed to become one of the major cities of the world. Such rapid development was not without its challenges and problems, however. Infrastructure in this frontier state was initially nonexistent and sanitary conditions were rudimentary. As late as the end of the 19th century, typhoid fever and other diseases plagued Illinois and other states in the Midwest. The annual typhoid death rate in Chicago in 1891 was 173 per 100,000 and 16 per 100,000 as late as 1910 -higher than in any city in the temperate zone. The highest death rate from typhoid, however, was in rural southern Illinois. Also, it was reported that "the state is not credited as recording as many as 90% of total deaths that occur within its limits" and "many typhoid deaths are not recorded" (Bartow, 1911, p.88). Waste from millions of people and animals were discharged directly into Lake Michigan, the Illinois River, and just about every other river in Illinois.

Founding of the Illinois State Water Survey

The Illinois State Water Survey (the Survey) was founded in September 1895 in Urbana as a unit of the University of Illinois Department of Chemistry. The Survey chemistry laboratory building was the only chemistry laboratory on campus at that time. In 1902, the Survey moved to the new Noyes Laboratory in Urbana. A plaque commemorating the pioneering work of Dr. Arthur W. Palmer, the first head of the Survey from 1895 to 1904, will be installed at the Noyes Laboratory this year. The Survey had its home in the Noyes Laboratory until 1951, when it moved to the newly constructed Water Resources Building in Champaign. In 1983, the Survey moved to its present home at the Water Survey Research Center on south campus. Regional offices are maintained in Peoria and Carbondale.

Palmer, a chemist, was meticulous and thorough. His detailed reports of water quality set a standard not only for the Survey, but also for the nation. His data for the Illinois River and the Mississippi River in the 1890s and at the start of the 20th century have been used, in conjunction with recent data, to establish century-long trends in nitrate concentrations in the Mississippi River basin as part of a national assessment of hypoxia in the Gulf of Mexico (CENR, 2000). Palmer also used these early data in a U.S. Supreme

Court case to resolve a dispute brought by St. Louis against Chicago after the Lake Michigan diversion was initiated in 1900. The diversion flushed sewage from Chicago and other towns down the Illinois River into the Mississippi River, from which St. Louis obtained much of its water supply. After 30 years of controversy with the lake states over the amount of water being diverted - in excess of 8,000 cubic feet per second (cfs) - the U.S. Supreme Court in 1930 limited the amount of water that Illinois could divert from Lake Michigan. The diversion controversy erupted again in the 1950s and 1960s, and the Survey played a major role in providing data and expertise to the Supreme Court, which set the diversion to 3,200 cfs. Today, some of the water diverted from Lake Michigan dilutes waste, supports barge transport on the Illinois River, and provides water to millions of residents in the Chicago Metropolitan Area. The Survey maintains a dense raingage network across the region to measure precipitation as one component of assessing the water diversion.

Palmer also was an early protagonist of interdisciplinary research, which is widely recognized today as a necessary step in addressing complex environmental problems. Specifically, he promoted chemistry, geology, bacteriology, and engineering research to address the many water quality problems in Illinois (Hays, 1980). Bacteriological and biological research was necessary in order to understand the links between water quality and microorganisms, which, for example, both fixed nitrogen in surface waters and removed nitrogen from surface waters and sediments. Palmer worked closely with Professor Stephen A. Forbes, who had succeeded John Wesley Powell as curator of the State's Natural History Society's museum in 1872. Forbes pioneered research on the biology and water quality of the Illinois River and set a solid foundation for ecological studies (Hays, 1980). "In later years, this pervading interest in the interactions of organisms earned Forbes recognition by many as the true "father of ecology" (Harlow B. Mills cited in Hays, 1980, p. 37).

Palmer was interested in solving problems, not just in studying them, and he recognized the importance of engineering as part of the solutions. Under Palmer's leadership, the Survey addressed water softening methods, sewage and wastewater treatment, water supply technology, and the establishment of sanitary standards for drinking water in the early years of the 20th century.

The original mission of the Survey was to survey the waters of Illinois to trace the spread of waterborne disease, particularly typhoid fever. By 1910, 21,715 water samples from 971 towns in all of the state's 102 counties had been tested. Many wells were condemned due to combinations of high bacteria levels, nutrients, and minerals: almost 80% of the wells less than 25 feet deep, 65% of those 25-50 feet deep, half of those 50-100 feet deep, and 20% of those over 100 feet deep (Bartow, 1911). "Every engineer employed by the State Water Survey since it was organized has had a part in the collection of data which appears in this bulletin, and every chemist at the Survey has had a part in the collection of data or in the analysis of waters" (Bartow, 1911, p. iii).

The Board of Natural Resources and Conservation

In 1917, the Scientific Surveys were transferred to the State Department of Registration and Education, and each head accepted the title "chief" (Hays, 1980). The Board of Natural Resources and Conservation (the Board), composed of eminent scientists and professionals selected by the Governor, was established to guide the activities of the Scientific Surveys and to ensure that the Scientific Surveys' work could be done without undue political influence.

Scientific investigations were expanded in the 1930s, and the state's first inventory of municipal groundwater supplies was published. Activities also focused on methods to determine water levels in wells, yield testing, and establishment of an ongoing survey of the state's surface water resources. This survey included a cooperative streamgaging program with the U.S. Geological Survey. In 1931, the Survey conducted a sediment survey of Lake Decatur - the first such survey ever made by a state agency (Gottschalk, 1952).

Professor Noyes served as Secretary to the Board for many years, thus helping to cement an enduring relationship between the Scientific Surveys and the University of Illinois at Urbana-Champaign. Eighty-five years later, the Board still governs the Scientific Surveys. The Board is chaired by the Director of the Illinois Department of Natural Resources and has representatives of the Presidents of the University of Illinois and Southern Illinois University, and experts in the scientific disciplines of the Scientific Surveys.

Recognizing the Importance of Water Resources Planning

Over the years, the work of the Survey has changed to meet changing needs while still providing some of the original services. The Public Service Laboratory, for example, continues to provide free analysis of water samples submitted by Illinois residents. The latest addition to the suite of analytes is arsenic, which was added in 2002 in response to the U.S. Environmental Protection Agency tightening of the arsenic drinking water standard.

By the 1930s, no authoritative data on groundwater supplies were available. Continuing the tradition of looking forward, Chief Arthur M. Buswell, who succeeded Dr. Edward Bartow in 1920, proposed "a comprehensive survey of the volume of groundwater available in Illinois" (Hays, 1980, p. 122). Twelve years later, Buswell broadened his proposal to include all the state's water resources and to estimate future demand. This program was "aimed at securing the information needed for the most beneficial development of the state's water resources" (Hays, 1980, p. 145). "Although this project was included in the budget requests for several years, it was not funded by the legislature"

(Hays, 1980, p. 203). Buswell also stressed water quality for industrial purposes, with particular emphasis on corrosion problems and newer water-softening methods (Buswell, 1952).

After the resignation of Arthur Buswell, whose leadership spanned nearly four decades, William C. Ackermann was appointed Chief in 1956. Ackermann, a civil engineer, had worked with the Tennessee Valley Authority's water control planning department before heading the Watershed Hydrology Section of the U.S. Department of Agriculture's Agricultural Research Service. Working closely with John C. Frye, the new Chief of the Illinois State Geological Survey, the two Chiefs opened a joint regional office in Naperville to work on groundwater studies in the Chicago Metropolitan Area (Hays, 1980). "But no consistent progress had been made toward comprehensive mapping of the groundwater resources of the state" (Hays, 1980, p. 164).

Returning to Washington, D.C. in 1963, on leave from the Survey, Ackermann became "the first technical assistant for water resources in the Office of the President" (Hays, 1980, p. 170). "He put together, among the federal agencies, the first coordinated water resources research budget, first coordinated water research plan in the federal government ..." (Donald Hornig, cited in Hays, 1980, pp. 170-171). Building on this experience, "Ackermann organized and served as first Chairman of the Committee on Water Resources Research, with members from various government agencies dealing with water resources... Of all the needs in national water policy, Ackermann found the need for more advanced planning in regard to water resources to be among the most pressing.... It also led to Ackermann's increased concern for better planning at the state level, ... In 1965, Illinois Governor Otto Kerner designated Ackermann as director of a task force to formulate a comprehensive state plan for water resources" (Hays, 1980, p.171). In 1967, an ambitious state water plan was released, but was not implemented due to budget constraints (Illinois Technical Advisory Committee, 1967).

Ackermann was very active in various activities of the American Society of Civil Engineers (ASCE). He served in various capacities and held several offices, including the Director of Zone III of ASCE. During his lengthy tenure as Chief of the Survey, he contributed significantly to water resources planning and management in the state and the nation.

Recognizing a central scientific role in the evolving environmental movement, Ackermann developed "a strong advisory and supporting role with the newly created state environmental agencies." "The Water Survey itself was now organized in five principal sections: water quality, chemistry, atmospheric sciences, hydrology, and hydraulic systems. Increasingly, investigations were crossing section lines" (Hays, 1980, p. 179).

Population growth in the late 1950s and 1960s created the need for expanded water resources throughout the state, and the Survey attempted to identify and increase usable supplies. Studies addressed reservoir development and maintenance, new methods for evaluating wells and aquifers, and the effects of future development. A statewide network of observation wells was established, and investigations of groundwater resources in the Chicago and East St. Louis areas were important contributions to an inventory of the state's principal groundwater formations.

Contributions in Hydraulics, Rivers, Streams, Lakes, Wetland Dynamics, and Groundwater Modeling

Over the last 70 years or more, the Survey's civil engineers have made significant contributions in the areas of open channel flow, sediment transport, groundwater modeling, and wetland dynamics. Many lakes in the nation, and especially in the central part of the country, are filling up rapidly with excessive sediment deposition. Since the original sedimentation survey of Lake Decatur, more than 170 lake sedimentation surveys have been completed by the Survey. These surveys, supported by yield analyses, have provided extremely valuable information to local communities and state agencies in predicting and managing their water supply availability and recreational opportunities.

One of the pioneering works on hydraulic geometries of streams and rivers was completed by John B. Stall and Dr. Yu Si Fok in the mid 1960s. Stall also served in various capacities for the ASCE, including a term as the Chairman of the old Hydraulics Division. The hydraulic geometry concept was further expanded by Dr. Nani Bhowmik and Stall to include floodplains. Dr. Ted Yang and Stall later developed the Unit Stream Power concept, which is used by various entities for modeling sand transport in streams and rivers. Stall also was instrumental in the development of the Illinois Urban Drainage Area Simulator model, which is used all over the world to evaluate and model urban runoff.

The 7-day, 10-year low flow concept for the determination of minimum flows and assimilative capacities of streams and rivers essentially was developed at the Survey by Stall and Dr. Krishan P. Singh. This concept is now used widely for managing low stream flows for a variety of purposes.

The Survey's contributions in river hydraulics are many and varied. Bhowmik was the first engineer in the nation to conduct a scientific evaluation of the physical impacts of navigation in large rivers, including the upper Mississippi, Illinois, and Ohio Rivers. These research activities started in the early 1980s and have provided valuable information for potential navigation improvements and ecosystem-based management of the upper Mississippi and Illinois Rivers. Bhowmik and Dr. Mike Demissie also worked on the Long Term Ecological Research of Large Rivers, a multidisciplinary endeavor by engineers, biologists, and geologists, supported by the National Science Foundation. The concept developed from this project formed the basis for the establishment of the Environmental Technical Management Center in Onalaska, Wisconsin, for research and data collection on the upper Mississippi River. Bhowmik contributed significantly to the understanding of waves generated by recreational boats in lakes and rivers. Also, he has served in various capacities for ASCE, including chairman of a Task Committee, control and contributing member of several Task Committees, and as an Associate Editor of the *Journal of Hydraulic Engineering.*

The Survey's contributions on the effects of wetlands in flood peak attenuation received national attention, including the deliberations of a Task Force appointed by President Bill Clinton during the 1993 flood on the upper Mississippi, Missouri and Illinois Rivers. That report prepared by Demissie and Dr. Abdul Khan was cited in numerous publications as a significant contribution to the effects of wetlands in flood peak attenuation for smaller watersheds. Contributions by Demissie and Bhowmik in the basic determination of sediment load transport and sedimentation of the Illinois River, Kankakee River, and other rivers and streams of Illinois have assisted and will continue to assist the state, and federal and local government in the technical evaluation and development of management alternatives for these streams and rivers for many years to come.

Contributions by Survey engineers in the modeling and evaluation of groundwater resources also are many and varied. Bill Walton's groundwater book has been a standard for many years. Tom Prickett and Carl Lonnquist prepared a classic groundwater modeling report for the Survey in the mid 1960s that has been a standard modeling report for numerous state, federal, international, and private entities worldwide.

Over the years, Survey engineers have played and continue to play a very active role in numerous ASCE activities. These included Director of Zone III (Ackermann), chairing of a Division (Stall), co-editing national conference proceedings (Bhowmik), chairing and participating in ASCE Task Committees (Bhowmik, Demissie, Stall, and Singh), chairing technical sessions, and presenting numerous papers at ASCE conferences.

The Emergence of Atmospheric Sciences and Long-Term Monitoring

During World War II, Survey chemists cooperated with the University of Illinois and the federal government in studies on the detection of chemical warfare agents in water and methods for their removal. Meteorological efforts expanded in the postwar years, including the investigation of atmospherics for cloud seeding to induce rainfall, use of radar to measure rainfall and track severe storms, and the establishment of networks of densely spaced raingages (Changnon and Huff, 1997). The U.S. Weather Bureau transferred the State Climatologist to the Survey, and computerization of the Survey's historical weather records began. In 1953, the Survey gained national prominence when "radar observers were able to view and record on radar photographs for the first time the development, growth, and partial disintegration of a severe central Illinois tornado" (Hays, 1980, p. 158). In 2001, the magnetron from that early radar was preserved and put on display at the Survey.

Periodic heavy storms and flooding always have been a concern in Illinois, and precipitation frequencies issued by the Survey provide an important basis for civil engineering design and planning. Precipitation has increased since about 1970, and the Survey has reanalyzed the data and issued modified precipitation frequencies. Major field studies conducted by the Survey 25 years ago under the leadership of Stanley A. Changnon, an atmospheric scientist who was appointed Chief of the Survey in 1980, identified that heat islands over large urban centers such as Chicago and St. Louis increase temperature and stimulate heavier precipitation.

In the 1970s, the issue of acid rain emerged as an important environmental concern and an energy-related problem. The chemical composition of the atmosphere became a major focus of activities at the Survey. The National Atmospheric Deposition Program (NADP) was established in 1978 to monitor the chemical composition of precipitation across the nation. For 24 years, the Survey has analyzed the chemistry of weekly samples from the NADP network, now numbering more than 240 sites nationwide. Since 1992, the Survey has performed similar measurements for 10 sites collecting daily samples. Results of the chemical analyses show that the concentrations and amounts of sulfate in precipitation have decreased substantially since enactment of the acid rain controls under the 1990 Clean Air Act Amendments. Since 1997, all the other components of NADP, including an emerging mercury deposition network, data storage and analysis, and education have been transferred to the Survey.

For about two decades, the Survey has monitored a number of environmental variables statewide under its Water and Atmospheric Resources Monitoring program: weather and climate, radiation, soil moisture, sediment concentrations, and groundwater levels. Long-term monitoring of these variables will continue. The State Water Plan Task Force uses these records, in combination with data on river discharge and reservoir levels, for water resources planning and management, especially during droughts and floods. Scientists throughout the world use the soil-moisture data to study terrestrial-atmospheric moisture fluxes and budgets. Wind and radiation data from the program provide a foundation for evaluating the potential for wind and solar energy in Illinois.

The Midwestern Regional Climate Center, housed at the Survey, is a cooperative program with the National Climatic Data Center of the National Oceanic and Atmospheric Administration and provides many climate services to residents throughout the Midwest. The State Climatologist, still housed at the Survey, provides data and information on Illinois weather and climate to the news media, government agencies, private companies, and Illinois residents.

The Survey Now and in the Future

Since 1995, the Scientific Surveys and the State Museum have been Divisions of the Illinois Department of Natural Resources. The Scientific Surveys remain affiliated agencies of the University of Illinois at Urbana/Champaign. Support for scientific programs at the Survey includes a state appropriation and income from grants and contracts with various Illinois state agencies, municipal groups, universities, private organizations and businesses, and many federal agencies. The Survey has a diverse, multidisciplinary staff of 200, including 25 Ph.Ds. The Survey's major activities continue to be resource monitoring, research, and services in five Sections: Watershed Science, Groundwater, Atmospheric Environment, the National Atmospheric Deposition Program, and Analytical Chemistry and Technology.

Following a tradition of success and innovation for more than a century, the Survey continues to adapt to changing priorities and implement new programs while providing traditional services, all within the framework of the hydrologic cycle.

Climate change and its possible influence on state water resources are important to the state, the nation, and the world. Unfortunately, precipitation scenarios for the Midwest for the 21st century range from an increase in mean annual precipitation of about 10 inches to a decrease by the same amount. Such uncertainty makes water resources planning very difficult. To reduce this uncertainty, Survey scientists are starting to conduct diagnostic studies of computer models which simulate the climate system in an attempt to understand the reasons for the divergent climate scenarios. They also are developing a new regional climate model embedded in a global climate model. Linked to the climate model is a regional air quality model and a hydrologic model. The results from this research could provide an improved scientific basis for making policy decisions about climate change, air quality, and water resources.

Working with the U.S. Army Corps of Engineers, the Survey has modeled hydrology and hydraulics to evaluate design options for constructing islands in the Illinois River near Peoria. Construction of these islands, announced by the Corps in June 2002, will remove large quantities of sediment from the Illinois River, help reduce sedimentation problems, and improve biological habitat. A related major project in partnership with the Corps is restoration of Illinois rivers. The Scientific Surveys are developing the Illinois Rivers Decision Support System to inventory all related activities and data, and to provide analytical tools for evaluating restoration options.

In many cases, it is difficult for small water suppliers in rural communities to keep pace with aging systems, new regulations, and increasing demands. To assist small water supply systems throughout the Midwest, the Survey, in cooperation with the Water Resources Center at the University of Illinois in Urbana/Champaign, heads the new Midwest Technology Assistance Center sponsored by the U.S. Environmental Protection Agency.

In summary, the following is a brief characterization of Survey history:

Era I: 1895 to1930: a primary focus on water chemistry
Era II: 1930s to 1960: a primary focus on hydrology (water quantity)
Era III: 1960 to 1990: an increasing focus on atmospheric resources
Era IV-present: a uniform focus on environmental issues in water resources and weather/climate.

Conclusions

For more than a century, the Survey has been a leader in water resources studies and in atmospheric studies for 60 years. There are many reasons for the agency's longevity and numerous successes. A significant reason is that generations of political leaders and constituents throughout Illinois, and especially in Springfield, have recognized the economic, environmental, and human health benefits to Illinois and the nation of investing in science. Based on extensive scientific data and analytical tools produced by the Survey, private individuals, government agencies, private companies, and nongovernmental organizations have made hundreds of thousands of decisions to protect human health, locate and protect water supplies, stimulate economic growth, protect the environment and ecosystems, and enact science-based policies and regulations. Other reasons include the diverse scientific backgrounds of the Chiefs and Survey staff and their ability to form partnerships with many national programs and professional organizations. The Survey has had only eight directors or chiefs in 108 years, thus providing a great deal of continuity and stability to the organization.

The Survey has achieved national and international recognition as a unique source of data and information on water and atmospheric resources. Survey scientists and staff participate in and lead national programs with benefits that far transcend state boundaries.

However, a major challenge that has faced Illinois and the Survey for 70 years still remains - developing and implementing a comprehensive survey of state water resources and a statewide water resources management plan. A key unresolved question, asked by L.R. Howson of Chicago in 1951 is "... to what extent that interest [state interest in its water resources] should be expressed in control legislation?" (Howson, 1952, p.145). Howson's view was that "... the regulation of many so-called water shortages in a state such as Illinois, which has abundant water resources, in many localities a choice of resources is quite largely an economic question" (Howson 1952, p.147).

Debate between a free-market approach based on the riparian doctrine of reasonable use and a regulatory approach to allocating water resources has been renewed in recent years. In April 2002, Governor George Ryan signed an Executive Order requiring the Department of Natural Resources to take the lead in preparing an integrated groundwater and surface-water resources agenda and assessment report, and to submit this report to the Interagency Coordinating Committee on Groundwater. In turn, the Committee shall report to the Governor by January 1, 2003, and establish a water-quantity planning and management procedure for Illinois.   Whatever the outcome, the Survey, working closely with the State Geological Survey and the Office of Water Resources, is committed to providing an improved scientific basis for water resources planning and management (ISWS, 2001). Currently, the Survey is undertaking a multi-year project with the State Geological Survey to characterize water resources in Kane County in the Chicago Metropolitan Area, and with the Mahomet Aquifer Consortium in central Illinois.

One of the greatest assets of the Survey is its collections of data and reports. It is a Survey goal to make all its major reports and data sets dating back to 1895 available via the Internet. Substantial progress already has been made in this endeavor (www.sws.uiuc.edu). By next year, the Survey also plans to have more than 300,000 groundwater well records digitized and available via the Internet.

Throughout its history, the Survey has enjoyed and benefitted from close professional ties with the American Society of Civil Engineers. The Survey salutes the American Society of Civil Engineers in celebrating its first 150 years of achievements. We look forward to continuing the productive partnership.

Acknowledgments

We thank Debra Mitchell and Patti Hill for administrative assistance, and Eva Kingston for editorial suggestions.

References

Bartow, E. (1911). *Chemical and Biological Survey of the Waters of Illinois*. University of Illinois Bulletin, 8(23), Urbana, IL.

Buswell, A.M. (1952). "The State Water Survey Division 1920- ." *Conference on Water Resources, October 1-3, 1951*, Illinois State Water Survey Bulletin 41, Urbana, IL, pp. 23-26.)

Committee on Environment and Natural Resources (2000). *Integrated Assessment of Hypoxia in the Northern Gulf of Mexico*. National Science and Technology Council, Committee on Environment and Natural Resources, Washington, DC.

Gottschalk, L.C. (1952). "Sedimentation Problems." *Conference on Water Resources, October 1-3, 1951*, Illinois State Water Survey Bulletin 41, Urbana, IL, p. 51.

Hays, R.G. (1980). *State Science in Illinois: the Scientific Surveys, 1850-1978*. Southern Illinois University Press for the Board of Natural Resources and Conservation of the Illinois Institute of Natural Resources, Carbondale, IL.

Howson, L.R. (1952). "State Water Policies." *Conference on Water Resources, October 1-3, 1951*, Illinois State Water Survey Bulletin 41, Urbana, IL, pp. 145-149.

Illinois State Water Survey (2001). *A Plan for Scientific Assessment of Water Supplies in Illinois*. Illinois State Water Survey Information/Educational Material 2001-03, Champaign, IL.

Illinois Technical Advisory Committee (1967). *Water for Illinois: a Plan of Action*. Illinois Technical Advisory Committee on Water Resources, Springfield, IL.

Water Quality in America since 1980:
From Nitrates & Non-point to Ecosystems, Toxics and Terrorists

David J. Allee[1]
Leonard B. Dworsky[2]

Abstract

Selected themes through the history of water quality management paint an exciting picture of accomplishment led by innovative and committed Civil Engineers. Research results on the tools for waste management and its analytical basis, for example, allowed institutional capacity to evolve to implement them. This process continues. Currently the field is facing a number of frustrating issues – the history of a few are summarized with some hints of how they might be resolved. For example, drinking water protection has proceeded apace, helped by a series of national initiatives and now the concern for bio-terrorism. But small, scattered systems, public or private, water supply or onsite waste disposal need attention in ways they have not in the past. Organizational challenges include who should pay. Finally we indicate some sources of further opportunity. For example, now that there is widely accepted methodology to put a dollar value on anything, when should it be used?

Introduction

We make special note of two journal papers both co-authored by Leonard B. Dworsky; "Water Resources Planning And Public Health; 1776-1976" ASCE, March 1979, and "Water Pollution Control", American Geophysical Union," January 1977. We dedicate this limited contribution to the history of civil engineering in "an appreciation to those that have gone before" and particularly to the memory of Professor Bernard B. Berger, the other co-author of the two papers that are the foundation for this one.

For the last several decades we have enjoyed the efforts of an annual multi-disciplinary seminar group to understand the public policy process in water resources and the environment. Of particular concern has been the evolution of public capacity to plan and manage ahead of the impact of growth. This has been built upon the following basic notions about the policy cycle which shape our view of history.

The policy cycle produces incremental changes. Concerns usually provoked by an event (a crisis?) lead to specification of an issue by a group. Alternatives that modify existing agency activities (information, bribes, coercion and restructuring of power) are placed on an agenda. Consequences (will it work and who cares) and modifications are debated. Chances are that the more the issue has been "incubated," such as through a participatory planning process, the larger the change from past programs that can be gotten through the many checks and balances in the system. A crisis with little

---

[1] Professor of Resource Economics, Department of Applied Economics and Management, Cornell University, Ithaca, NY 14853; phone 607-255-6550; dja1@cornell.edu

[2] Professor Emeritus of Civil and Environmental Engineering, Cornell University, Ithaca, NY 14853; phone 607-257-4153; lbd3@cornell.edu

preparation before it occurs will achieve less response than would be expected from the severity of the crisis. With many veto points passed and working majorities achieved, implementation can proceed. The more severe the crisis the less political capital needed to get agreement.

Implementation often involves a changed set of participants than during policy formation and at least that will produce a different result than early participants expected. Compromises and imperfect science will almost assure results that don't quite satisfy those that were originally concerned; thus, with evaluation (mostly informal) the cycle will be repeated. But now the cycle proceeds with the results of the above changes available and perhaps in response to a new crisis. A string of small increments makes knowing what policy is at any one moment challenging.

Budgets reflect a lagged sense of priorities as well as the inequalities in political power between stake holders and other factors in society - such as acceptance of governmental roles. None-the-less public budget decisions say a lot about priorities and changes in priorities. Both federal funds for enforcement and investment in water quality infrastructure, in terms of purchasing power have declined, while those for more direct health services have increased dramatically. Some argue that environmental protection is inherently regulatory in nature and thus has less potential to affect budget levels than other public concerns such as the many parts of income equity or concern for development.

Environmental problems ("tragedy of the commons"), for example, have faced governments since the first village, but they have always had to overcome the fact that the benefits while large in aggregate are diffused over the many, while their costs are concentrated on the few. Thus values and constitutional protections seem to play such a large role compared to analysis based on hard science. A particular challenge to engineering whose tools are to apply that science, thus the importance of growing institutions that can support those tools.

How is the nation organized to manage water resources and other aspects of the environment? What technologies are we trying to put in place? A new desirable technology creates a gap between what we are achieving and what we could achieve. This gap provides an incentive for change. Wider understanding of non-point sources is a current case in point. But to carry out the shift to new technologies and the use of new knowledge old institutions must be displaced or modified. Hence the expansion of programs directed toward stream banks, storm water and farm practices among many others. A feel for all of this can be gotten from looking at the way different agencies have fared in the competition for authority, people and budget. This is usually the result of the interaction of organized and varied interests interacting with legislators and the agencies, sometimes called the "iron triangle", but more accurately termed as the power cluster activated by the alternatives and consequences stimulated by the policy cycle.

Prologue

Selected themes for a historical perspective before 1980 should aid the reader in assessing the Nation's approach in more recent years.

Theme 1: Ever Broadening Technical Scope. Sanitary Engineering and Sanitary Science must be used here in their broadest and most generic sense. While "Sanitary Engineer" was the formal and legal title of commissioned officer engineering personnel in the Public Health Service until July 1980, the engineering disciplines represented by that title ranged widely. Similarly, the Sanitary Sciences encompassed all the basic areas needed as the tasks broadened with time. All these, together with Medical and other Officers of the Public Health Service were Sanitarians in the broadest sense as specialists in Sanitary Science and Public Health. It was this collective team that contributed substantially to the policy foundations of modern environmental management, to the point that the use of the title "environmental" engineering is common place. We speak to the use of all of these skill and knowledge sets and expect the future to add more, as we are adding habitat specialists and ecology now.

Water is a main component of human life. As the conditions and values of human life change, so must the management of water resources change. Thus, policies controlling water resources must continually be tuned to serve a changing scene.

Theme 2: Building the Basic Tools. In 1913-1917 under the direction of Dr. Wade Hampton Frost, a new group of medical officers, engineers and scientists were located in an abandoned Marine Hospital in Cincinnati, Ohio. They poured forth much of the fundamental information that laid the basis for the control of water pollution:

Streeter and Phelps define the Oxygen Sag Equation
Theriault confirms the rate of Oxygenation of Polluted Waters
Streeter confirms the rate of Atmospheric Reareration
Hoskins outlines the elements of bacterial pollution
Purdy develops major elements of stream biology, and
Phelps and Hommon initiate studies on industrial wastes

The more recent characterizations of Giardia and Crytosperidium, or new understanding of pesticides and other toxics, nutrients and many, many other "breakthroughs," too numerous to list have driven institutional change and will continue to do so!

Theme 3: Links to Basin Planning. The first national report with a title like "Water Pollution In The United States:" was issued 1939 by the National Resources Committee as House Document 155, 76th Congress, 1st Session. A demonstration river basin control program, initiated by the National Resources Committee for the Ohio River Basin was reported out in 1943 as Ohio River Report, House Document 299, 78th Congress, 1st Session. This joint report of the Public Health Service and the Army Corps of Engineers became the classic study. Many comprehensive basin plans, but none prepared after 1980, contained detailed sections on water quality. Some of the few multi-purpose compact basin commissions continue to do so, and a number of basins have enjoyed federal support for a new version of the old water quality conference led by USEPA with multiple agency funding but usually not including integration with tradition quantity development projects. The traditional federal development agencies or their client organizations are increasingly involved for other aspects such as habitat restoration and non-point source management.

Theme 4: National Study Commissions. The National Water Commission was created by the Congress September 26, 1968. It conducted a five -year study of national water resources problems and programs and reported to the Congress and the President on its findings and recommendations. This was one in a series of such increasingly comprehensive commission reports extending back to Teddy Roosevelt that appeared about every two decades. A summary of 22 General Findings and Conclusions was presented by Theodore M. Schad, Staff Director, at the American Society of Civil Engineers National Water Resources Conference, Washington DC. January 31,1973.

Seven major ideas provided the foundations for the Commission's recommendations. These included:

1. Demand for water in the future is not pre-determined, but depends largely on policy decisions that can be controlled by society;
2. Future water programs should shift emphasis from water development to reservation and enhancement of water quality;
3. Planning for water development should be linked to water quality and coordinated with land use planning (which is a local, not a national or state unction);
4. More efficient use of water in agriculture, industry, and for domestic and municipal purposes is essential to reduce waste;
5. Sound economic principles must be adopted to encourage better use of water resources.
6. Updated laws and institutions are needed if future water policies are to be successfully implemented
7. Development, management, and protection of water resources should be controlled by the level of government (federal, state, local or regional) that is closest to specific problems and capable of fairly representing all classes involved.

As explained by Commission Chairman Charles Luce, the comments and findings of the Commission and were exemplary of the purposes of the ASCE report entitled"1776-1976". Note this was followed by a National Water Quality Study Commission chaired by then Vice President Nelson Rockefeller who while governor of New York in 1960 pushed through the first billion dollar funding for sewers by any government. In historic perspective, the Act that authorized a federal water pollution control program wasn't called the "Clean Water Act" until it was amended in 1977 in response to the "mid-course corrections" in response to the recommendations of the Study Commission.

The "Western Water Policy Review Advisory Commission (WWPRAC)" Act of 1992,sponsored by Senator Hatfield (OR) had initially reflected his intent to have a national policy review. Subsequent efforts on his part to so amend it did not materialize. WWPRAC was established to undertake its comprehensive review of federal activities in the nineteen western states (the traditional seventeen plus Alaska and Hawaii) which directly or indirectly affect the allocation and use of all water resources. It was to submit a report of findings and recommendations to the Congressional Committee's having jurisdiction over federal water programs and who were represented on the Commission. Conflict within the Commission and with the key Congressional Committee leadership over the impacts and significance of irrigation water use kept the results from being formally accepted. Note that a comparison between Committee website material and the Commission recommendations by one of our seminar teams suggested close overall correlation in topics and opinions. Ironically, this speaks to the value of such studies in providing "fresh" summaries of issues and options to incubate those issues for the succeeding policy cycles. Since our testimony in favor of the establishment of the WWPRAC, we followed with great interest Commission process, procedures and products. We are pleased to report that all three were carried out in an exemplary manner.

Will a constituency for such policy reviews appear on schedule again in another decade or so? If so, will it be driven by the views of civil and environmental engineers? Given the historical track record of adaptation and function we have to say, certainly! But the company they keep will continue to reflect the trend of increasing complexity in issues and stakeholders.

Theme 5: Comprehensive Research Programming. The report "Goals Of Civil Engineering Research, Its Responsiveness To The Needs, Desires, An Aspirations Of Man" was by ASCE's Committee on Research in cooperation with Rose-Hulman Institute Of Technology, Terre Haute, Indiana, Sept 1-4.1971. Three basic types of research and development were stressed;

(1)   enhancing the engineers role in management.
(2)   improving the technical capacity for designing and constructing environmentally and economically sound engineering works.
(3)   identifying, as quantitatively as possible, the side effects of engineering works on ecology, quality of life, and the overall well-being of society and its surroundings.

A special need for improved communication in situations requiring expertise in fields other than engineering was recognized. This was followed by a set of improved water resources research activities in the early 1970's:

> Establishing the Committee on Water Resources Research of the Federal Council for Science and Technology within the Executive Office of the President.

> Leading to a ten year program of Water Resources Research.

> Water Resources Research Centers at Land Grant Colleges.
> facilitating and encouraging social science research (politics, economics, law, sociology) along with the life, physical sciences and engineering, in support of joint projects.

> Stimulating information exchange through National Water Resources Scientific Information Center.

These activities have not sustained the purchasing power of the funding support that they attracted in these early years. Research support for water quality has done better through specialized water quality oriented programs most administered by USEPA such as the STAR program, or farm oriented research through the USDA/Land Grant System such as in support of best management practices for non-point sources.

In a review of the federal Water Resources Research program in 1981, as required by the 1978 Act, the National Research Council noted, "As a consequence of the abolition of the Federal Committee On Water Resources Research, no interagency committee existed to undertake the coordination functions of the 1978 Act. ... The deficiencies noted ... are convincing evidence that the ad hoc approach to management of the federal water resources research program will not yield the results expected by Congress when it enacted the Water Research And Development Act Of 1978." This 1978 Act re-instituted what had started with the corresponding Act of 1964. Then in February 1984 the Congress of the United States again re-enacted the Water Resources Research Act (PL. 98-242) which was vetoed by President Reagan. On March 22, 1984 the Congress overrode the veto by a very wide margin in part because of the way the White House surprised them with the veto and certainly also because the topic was seen as close to "motherhood."

The strongest continuous support for the Water Resources Research Program and the research centers at the land grant colleges has come from within the Congress. Each member seems aware of the importance of water and the need to seek new knowledge to help in developing solutions. It is clear, too, that the availability of research capacity in each region and state and territory of the Nation on a decentralized basis is considered to be a contribution to an effective Federalism, which does not have to depend solely on Federal entities for necessary services. In "Envisioning the Agenda for Water Resources Research in the Twenty-first Century," a new report by the National Research Council Water Science and Technology Board has 43 recommendations, cast broadly into three categories: water availability, water use, and water institutions. Perhaps this can be built into further action by the Congress.

Theme 6: Policies For The Future - Revolution Or "Déjà vu all over again." Water quality has been the focus for significant changes, approved by the Congress and confirmed in the courts, to greatly strengthen the relation of the federal government to the states and to local governments. Particularly true in 1948 and 1972, but also with many amendments in between (at least 1956, 1961, 1965, 1966, 1967) and since. Most important, the Congress declared in 1972," The objective of the Act is to restore and maintain the chemical, physical and biological integrity of the Nations waters." Secondarily, it established two additional goals; the discharge of pollutants into the navigable waters of the United States be eliminated ... And that wherever attainable, - an interim goal of secondary treatment - be achieved … (emphasis added).

Over the past several decades responsibility for water and natural and environmental resources fluctuated between national and state and local governments. Some have argued that the programs for big technological investments have done their job and now more life style changes are need to make further gains in environmental integrity. Shifts in institutional arrangements began to change substantially in the 1980's during the Reagan administration. The 1994 mid-term elections sharpened the debate over the allocation of functions. As the Nation moved toward 1996 it was important to understand the fundamental nature of the long ongoing debate about the Federal system. Should the federal role grow, if so where and how? If the states lag, and competitive relations are affected, whose responsibility is it to prod them? What is the role for direct federal/local links? And many more questions of a similar nature are reflected in the evolution of water policy.

It is not a simple question of which party is in power. President Nixon signed the National Environmental Policy Act of 1969. In the 1981 President Reagan terminated the implementation of the Water Resources Planning Act of 1965, negating years of experience under that format. Initially enacted with strong bi-partisan support, the 1965 Act was intended to be the vehicle to update the Nation's water resource posture and policies. New laws and the courts changed water allocation processes, increasing the need for interstate coordination. The national and international concern for environmental matters confronted the American public with new problems and issues. Boundary water problems with Mexico and Canada become more intense and, compounded by the free trade agreement. And history and the courts have added new members to the Federal System; the governments of the numerous Native American communities.

So how have we coordinated and integrated water investments and regulations if not by comprehensive basin planning and federal level institutions like the Water Resources Council as called for in the 1965 Act? Planning by default appears to have shifted away from, not closer to, an ideal to apply rational analysis. The nation appears less well organized to apply analysis across all water resources, across all uses, taking responsibility for all alternatives and all consequences even in a planning mode that was not expected to go beyond recommendations to the Congress, the President and the States. After all we in the water game understood that after the attempt on the Missouri River failed a second Tennessee Valley Authority was not in the cards even though TVA's survival strategy of serving as planning staff support to the governor's in the region has worked.

Note that hydrologic and hydraulic models had become much more user friendly. For this reason and the greater effectiveness of environmental groups at every level, down stream interests now have to be bargained with before a project is proposed. Water quality and environmental impacts are not left behind. Litigation and the last resort – play the Endangered Species Act card – also seem to have pushed the arena for projects back to the very local level. Political gains from multi purpose, multi objective forms of interagency collaboration seem to have dried up.

It is not clear how to bring an end to this comprehensive planning drought but we have proposed re-establishing the interagency coordinating committees so that at least interagency communication can be improved. The WWPRAC proposed a basin financing arrangements much like some enjoyed the Pacific Northwest but the ideas so far have found little traction. The WWPRAC also urged more attention to coordination at the watershed level and pointed to many good examples in the West most with federal agency participation. We agreed and cited many good examples in New York State. Federal agencies are less involved in the East. We also pointed out that such watershed groups might need a forum at the basin level that could be provided by an interagency coordinating committee hopefully without the specter of independent federal power raised by TVA or the 1965 Planning Act and by the WWPRAC basin funding entity. This might have more chance of success than building upon the basin and estuary interagency efforts led by USEPA of which the Chesapeake was the first and most successful, perhaps because of its evolution of a strong state role.

Major Issues

We provide guidance into the "New Century" by an assessment of issues that are frustrating us now.

Issue 1: Water Pollution Control. Contrary to the teachings of the 1980's that government is the problem, not the solution, the national water pollution control program stands as a positive counterpoint. First, the control of waterborne disease and second, the growing concern for safeguarding water quality for all human purposes and living things stand high among any of the accomplishments of any modern societies.
In retrospect, it is clear that the earlier, and later Clean Water Acts through the 1980's, were major forces in changing the nations attitude toward water pollution. In many ways it is a wonder that the nation has done as well as it has. Comprehensive planning to bring rationality to the billions of dollars for water pollution control was never used by the Congress. Much like the nation did when it ignored John Wesley Powell's admonition to save our scarce water to irrigate the best land, State priorities were most often set by the readiness of polluters. Suggestions for the integration of water pollution control with water resource development acts of the Federal agencies were seldom honored. Congress seldom concerned themselves with pollution control planning or with the large resource development plans of the nation. Prof. Lowi's definition of distributive politics was the course followed (Everybody gets a fair share of the public money). Yet it may be that in the "real world" of democratic governance such was all that we should expect. Perhaps the results were not as bad as one might think, not having tried the other options.

Americans need to understand that there is no end to the process in which they have now been engaged for a half-century since the 1948 Act. At some point the cost of the still current (and physically and biologically impossible) policy of "eliminating the discharge of pollutants to the waters of the nation" need to be confronted in the light of other challenges that must be met. "How clean is clean" inherently needs continuous determination, and it is not a technical question. American culture, social equity, and the meaning of the exponential curve of disturbance of the environment due to growth in population and income during the next quarter century must be confronted. A curve that represents understanding of the environment stimulated by the concern for the destruction of those resources must be accelerated and a corresponding curve reflecting the evolution of management capacity must be accelerated as well. We need to look hard at this evolving future. Much is going to depend on how it is interpreted.

Construction Grants Overview: During the 1970s and 1980s, the federal Construction Grants Program was a major source of funds, providing more than $60 billion for the construction of public wastewater treatment projects. These projects, which constituted a significant contribution to the nation's water infrastructure, included sewage treatment plants, pumping stations, collection and intercept sewers, rehabilitation of sewage systems, and the control of combined sewer overflows. With the 1987 amendments to the Clean Water Act, Congress established 1990 as the last year that construction grants funding would be appropriated calling for revolving loan funds in each state. But the picture is changing as new concerns and new goals are being added to the old.

In Washington, February 15, 2001 Senator Bob Graham, D-Florida, chairman of the Subcommittee on Fisheries, Wildlife and Water, introduced landmark legislation to ensure the environmental and financial sustainability of our nation's water programs. The legislation authorizes $35 billion over 5 years to modernize the operation of state water pollution control revolving funds. It also revises the allocation of those funds to ensure that funding priorities reflect water quality needs. Graham said, "Sections of the country that had an abundance of water are now looking at restrictions. This legislation moves us toward suggestions of ways in which the supplies we have can be used more effectively and new supplies can be developed. The bill represents a recognition that the federal government will be working with states and localities to accomplish these goals." These goals are now before the new Administration, and in a review of the supporting materials we note the integration of quantity concerns in the language almost to the exclusion of the quality protection results the funds being authorized were set up to achieve.

Issue 2: Safe Drinking Water – Toxics and Ecosystems? The Safe Drinking Water Act (SDWA), which celebrated its 25th anniversary in 1999, is the main federal law that ensures the quality of Americans' drinking water. Under SDWA, USEPA sets standards for drinking water quality and oversees the states, localities, and water suppliers who implement those standards. Testimony up to the 1974 passage was sharply critical of the Congress when it acknowledged that public water supplies had not had Congressional public hearings for over 40 years. While Congressmen Rogers and Magnuson. led the way, Congressman Howard Robison of New York, whose district included Cornell University, was given the honor to introduce the Act in the Hearings before the Congress as early as 1970. This allowed us to play a supporting role. Even prior to that effort President Lyndon Johnson sought a specific bill in 1968.

The official Drinking Water Standards were authorized in 1914 for the US Treasury Department. It was emphasized that they did not represent the nearest approximation to the purity which is desirable to attain in drinking waters, but rather the furthest deviation from purity considered permissible and safe. Standards were then established in 1925 and 1946. Later, these standards become the charge of the PHS, the Bureau of Water Hygiene, and eventually became a matter for water management by USEPA for water chemistry, biology and toxic properties. The explosion of chemicals in the environment and the growth in public expectations has put the standard setting process under substantial stress. On August 6, 1996, President Clinton signed Amendments to SDWA. The 1996 Amendments emphasize sound science and risk-based standard setting, small water supply system flexibility and technical assistance, community-empowered source water assessment and protection, public right-to-know, and water system infrastructure assistance through a multi-billion-dollar state revolving loan fund. Recently USEPA estimated that $150 billion will be needed over the next two decades by 55,000 community water systems. And in turn the General Accounting Office has urged that oversight tools be strengthened to improve such estimates and better identify common and recurring problems.

Since 1996, statutory requirements in the Safe Drinking Water Act included amendments from 1997, 1998 through 2002. A number of these were presented for ASCE action. They represent continuing Congressional interest. Some examples – 1997 USEPA Report to Congress called for Drinking Water Infrastructure Needs Survey (including Indian Tribes). In addition there was a call to develop a plan for additional research on cancer risks from exposure to low levels of arsenic. January 1, 1998 - States to submit to USEPA first [annual] compliance report. Publish a list of contaminants not subject to any proposed or final national primary drinking water regulation (must include sulfate). August 6, 1999 - Establish National Contaminant Occurrence Data Base. Promulgate final regulation establishing criteria for a monitoring program for unregulated contaminants. .January 1, 2000 Directed to propose standard for arsenic. The policy cycle effectively at work! But does the pace of change need to be accelerated? "9/11," and decentralization provide opportunities but so do concerns for toxics and system reporting.

We have followed the progress of environmental management organizations at both US borders as bellwethers of the policy system. On the Canadian side the International Joint Commission (IJC) has been an innovator. For example it took the use of ecosystem concepts seriously, ahead of agencies in either country and played a part in their absorption into the cultures of both. We were happy to play a part in that process through our seminar and on IJC advisory committees. In responding to references from the member governments and in its assigned role as monitor of the bi-lateral water quality agreement the IJC has repeatedly responded with advice on the handling of toxics. For example it lists specific toxics that should be removed from use and the environment completely. Is this a cutting edge of a larger response to the vastly different risk perceptions held by experts and professionals on the one hand and the general population on the other hand? Reporting rule changes for water purveyors seem to have reflected an attempt at a compromise between these positions. Planners and experts have been assaulted for at least 50 years for inadequate techniques that make better use of local information and local informants, and that help those who must support the results feel like they also own the process. We have made a lot of progress in such democratization, but we suspect the nation must and will do a lot more!

Issue 3: Protecting the Homeland. The new focus on homeland security since the attacks on the World Trade Center and the Pentagon on 9/11/02, the need to protect the U.S. against attack from without and from within, cuts across all sectors. Any new crisis offers opportunities to achieve old goals that have affinity with the new goals. Water resources are no different, particularly the long standing concern for protecting drinking water supplies. USEPA Administrator Whitman, in less than a year after, was able to announce that $53 million in planning funds was being dispersed to large systems to promote security in the 168,000 facilities. Representative of small systems urged similar funding. And the nation has the advantage that linking security of water to security of the nation has already achieved considerable attention.

Internet searches by our students during the opening year of this new century had little difficulty identifying challenging foci for strategic assessments of public concern. The United Nation's, China, Globalization, Nanotechnology, Cyber-space are examples. The intent was to assess these signals as a new strategic contribution, given the growing intensity of interest in the environment. A special report of a National Commission, "National Security Challenges" in the 1990's provided the rationale underpinning this assessment. It illustrates that the United States government was worrying about the environment in the new century in terms of national security.

Further, on March 1, 2001, Recommendations of the Hart-Rudman National Security Report said, "The harsh fact is that the U.S. need for the highest quality human capital in science, mathematics, and engineering is not being met." (U.S. Commission on National Security for the 21st Century) In the new venue, security funding and homeland security are seen as a source of both great strength and considerable vulnerability. The President recognized that infrastructure is essential not only for making our nation more prosperous but for making our homeland more secure. But the key to success in any set of goals, as we have charted in water pollution and environment, is the human capital prepared to help shape the social capital needed.

The enormous federal investment in water resources representing billions of dollars of federal spending could be endangered by continued fragmentation of the effectiveness and efficiency of government programs without a comprehensive water-quality, quantity, lands program to protect the homeland. The integrity of the ecosystems of the nation have long been seen as perhaps only sustainable if there are strong authorities charged with their protection. Will homeland security be the impetus that brings about such changes in the Federal System of the United States? Wars have had such impacts before but usually more temporarily than permanently. Or is central authority the wrong place to look for success? Perhaps more is to be gained from process and particularly information improvements?

The Presidents homeland security initiatives call for effective use of information and closer coordination across all levels of government to help resolve many of the disparate program conflicts for water as well as other needs. For example, water resources information is maintained, developed and used at the Federal, State, and local government levels as well as within the private sector. Such programs to use information, science, investment decisions, technology and public health, it is said, need to be formulated and developed both horizontally (among federal agencies and departments) and vertically (among the federal, state and local governments). This initiative could be a key component of homeland security by improving the way agencies work together to serve citizens and maximize benefits of the federal government's overall investment in water resources.

This is a national struggle. Federal, State, and local government agencies, as well as the private sector, must work seamlessly together. Having the right system of management, participation, communication - content, process, and infrastructure - is critical to bridging the existing gaps between the federal, state, and local governments, as well as the private sector. In particular secure information systems to streamline the dissemination of critical homeland security information and assure improved responses. Note that a second paragraph in the President's statement identifies the investments that need protection - river and harbor, dams, drinking water and special consideration for bioterrorism protection – the technical base for which is the history of sanitary engineering and the sanitary sciences.

Issue 4: Ecosystems and Watershed Planning and Management. An ecosystem is defined as place having unique physical features encompassing air, water, and land, and habitats supporting plant and animal life. USEPA supports some environmental planning that addresses all the factors, both natural and human, affecting an ecosystems of a given region. An example of protecting an ecosystem is the watershed approach in which all pollution sources and habitat conditions in a watershed are considered in developing strategies for restoring and maintaining a healthy ecosystem. This is also reflected in the general concern for "place based" projects. USEPA seeks to protect ecosystems that support plant, animal, and aquatic life through a combination of regulatory and voluntary programs designed to reduce the amount of pollutants entering their environment. Multi agency opportunities are many and complex.

Ecological risk assessment methods in the Wildlife Exposure Factors Handbook is an example. It provides data, references, and guidance for conducting exposure assessments for wildlife species exposed to toxic chemicals in their environment. It is the product of a joint effort by EPA's Office of Research and Development, Solid Waste and Emergency Response, and Water. The handbook promotes and foster a consistent approach to wildlife exposure and risk assessments, and increase the accessibility of the literature applicable to these assessments. Such a work drew upon the expertise of many other institutions.

Part of watershed history has been forgotten - the six decades of experimentation and trial and error by the engineers, scientists and medical personnel of the United States Public Health Service (PHS). From the start they were charged with investigating "..conditions influencing ... sanitation and sewage and the pollution ... of navigable streams and lakes of he United States." Their service should not be forgotten. They provided much of the knowledge base for including quality aspects in later studies of watersheds. As we have observed, the 1939 report of the National Resources Board kicked off many of the concepts of subsequent programs. Watershed work by the PHS was no exception. Many landmarks dot this landscape. Mentioned above is the Ohio River report with the Corps of Engineers, to be followed by an interagency effort on the Missouri, the Snake, Willamette, Illinois, Detroit and Lake Erie, among others. Other outputs included 225 sub-basin reports published in 1952, and with over 60 more published in the 1960's. These were funded and prepared under Sections 1 and 2a of the Federal Water Pollution Control Act of 1948 as amended which called for these watershed plans.

Plan implementation did not follow, perhaps a lesson for current efforts. The Congress did not find a basis on which to put these reports to use. Perhaps this was lost under the momentum of the trend toward facility funding through the states rather than directly as in quantity oriented development projects. Also the states expressed more than a little concern at the micro management of the PHS in directives for setting stream standards based upon intended use designations which was seen as a start on the process of directing how watershed were to be planned. The watershed effort failed to make the transition from PHS to Interior and then to independent agency status. Under Section 208 of 72-500, Area Water Quality Plans a similar "contract thinking approach" prevented that authority evolving into a watershed approach. Again implementation failed. In 1994 in testimony before the Senate on S. 1114 where Title III addressed Watershed Planning and Non-point Pollution we urged that a less structured, more facilitational approach be specified that would lead to project proposals. Unfortunately the whole watershed emphasis was a casualty of the conference committee process. Watersheds in USEPA have been "reinvented" with emphasis on the highly biologically productive and popularly supported estuaries and wetlands, as a perusal of

EPA web sites will show. More should be done to understand the viability of the multi jurisdictional entities involved.

A wider review, beyond the space allowed for this paper, would have included the outstanding work of the Forest and Natural Resources Conservation Services in the USDA and of the Bureau's of Land Management and Reclamation as well as other water agencies of state and federal governments to name a few, and the schools of agriculture, forestry and natural resources that have played important roles in both water quality and water quantity management.

Issue 5: Total Maximum Daily Load Program. USEPA is taking steps to achieve more holistic environmental planning and management by revising and reenvigorating the Total Maximum Daily Load - or TMDL—program. Established by section 303(d) of the Clean Water Act of 1972 and linked to Section 305b and dormant since passage, the primary mission of the TMDL program is to protect public health and ensure healthy watersheds by calculating safe and reasonable loading level for each pollutant and insuring that no more than those levels are released to the aquatic environment. It has received more attention in recent years as the limits of improvement by conventional investments and traditional programming come closer. As such changes often are, it was shaped by a strategy of law suites from various stakeholders.

The system of characterizing impaired waters has long been criticized for inconsistency over time and place to the point that comparisons can't be made. Are the problems getting better, and where, have been elusive questions. A February 2002 General Accounting Office review notes that some of the approaches have no appropriate scientific basis. Thus estimates of how many stream segments and water bodies need this attention (USEPA estimates 20,000) will have a margin of error. The proposed regulatory changes can be found in the August 23, 1999 Federal Register, and on EPA's TMLD website. But the most significant aspect of this change would be the production of a current plan for each of those affected water bodies an thus that many more opportunities for organizational capacity to grow apace to assure implementation.
Failure of most states to attain water quality goals and the pressure to extend regulatory measures to more sources, to reallocate existing load restrictions in response to long standing violations of ambient standards and the apparent cost benefit advantages of going to previously non affected sources may be too much to resist. But concerns include federal intrusion into land use controls, albeit indirectly through documentation and promulgation of load shares by the states. This may set up the basis for load trading, long advocated by economists. And the call for more accurate watershed based models and data may indeed be heard as far as the several appropriations committees. But once accepted planning methods are solved the creation of institutional arrangements to manage watersheds should not be far behind. Since this will involve trading, whether compensated or not, between point and non-point sources the institutional competence and trust required may exceed anything the nation has seen in watershed multi jurisdictional organizations. But it is perhaps not beyond the achievements of other special, local, public districts, a favorite form of such public management.

Issue 6: Non-Point Source Pollution. Congress amended the Clean Water Act (CWA) in 1987 to establish Section 319, The Non-point Source Management Program to help focus State and local non-point source efforts. Under section 319, State, Territories, and Indian Tribes receive grant money which support a wide variety of activities including technical assistance, financial assistance, education, training, technology transfer, demonstration projects, and monitoring to assess the success of specific non-point source implementation projects. Thus there are 15 years of experience dealing with non-point sources upon which the implementation of programs like TMDL and watersheds can be based.

USEPA offers as typical the Spragues Cove Remediation Project which in June 1995, Marion, Massachusetts, completed construction of a wetlands system designed to reduce storm water pollutant discharges. A wetlands system was chosen from a variety of technology concepts for its efficiency and efficacy. Elevated levels of fecal coliform bacteria were the primary concern. They had contributed to the closure of shellfish beds in the Cove and threatened nearby swimming beaches. To obtain funding for the remediation structure, the town joined the Buzzards Bay Project of the National Estuary Program in competing for a "section 319" grant administered through the Massachusetts Department of Environmental Protection, Office of Watershed Management. The town also received grant monies from the U.S. Fish and Wildlife Service and the Marion Cove Trust. Once grant funding had been obtained, the Buzzards Bay Project requested technical assistance from the USDA Natural Resources Conservation Service (NRCS). NRCS put together an interdisciplinary team of engineers, biologists, soil conservationists, a geologist, and a soil scientist to work with the town and the Buzzards Bay Project. This team helped the partners identify alternatives and select best management practices. Would this process lead to even more effective results if more watersheds were organized and had the TMDL data and analysis available?

Issue 7: No New Supplies Except Reuse As we noted the WWPRAC report of 1995 was met with antipathy by the Congressional Committee leadership – it had called into question the feasibility of so much western water staying in irrigation use when other uses were seeking supplies. The Commission laid its claim on the convergence of a number of trends, of which two are paramount. The first is the staggering population growth projections that are made for the 21st century. If present trends continue, the 2020 population of the West may increase by more than 30 percent. The second is the Congress's finding that current federal water policy suffers from unclear and conflicting goals implemented by a maze of agencies and programs. The Commission says that this finding was reinforced and documented by its investigations.

A third characteristic that makes the previous two of real world significance is the fact that the present western water resource base has been exploited to a high degree: that the future cannot depend on new water supplies to meet the projected water demands, and that answers must be found in the reuse, reallocation, redistribution, conservation and more efficient and equitable management of the current available water resources above and below the ground. The separation of water supply and waste water planning and management will be under intense pressure. We note that USEPA is considering the merger of the Clean Water and Drinking Water Programs because of the value of solving problems at the same time with particular emphasis on TMDLs to protect supplies and on a merger of standards.

The time has come when this conclusion must be believed, that there is not enough new water in the West to provide for the future of its citizens and their social system, including the non-human living things. Given the growing value placed on in stream values, the East is not far behind. Some program consolidations may be feasible and helpful, but they are no substitute for the information and organizational effort needed to make democracy work better, bottom up as well as top down.

Challenges

We have reviewed some basic themes in the pre 1980 history of water quality and drinking water protection and touched on the history of some current stumbling blocks to progress. Now we turn to some opportunities that are less clear, more ambiguous in our view.

Responding to" One Size Fits All?" National standards for minimums in delivered water and source water have been set contaminant by contaminant, sometimes for families of contaminants. Ambient standards have been tailored to intended use in both measurable and descriptive terms, albeit sometimes vague incomplete and not achievable. Other threats to health and life, to species retention, are set one by one. But they all come together in a community with limited social capital and other resources. Who sets priorities and how? Who decides that priorities have to be set and how, ie., that in some time frame they are not attainable? Are some standards so expensive to attain in some places that the effect on the goals for health, on life expectancy, on species retention, should be met some other way EPA's program for the support of "Place Based" initiatives is one of the responses to these concerns. The challenge to the analysis of the tradeoffs implied in terms of our knowledge and willingness to pay for the analysis is only secondary to the paucity of trust that we have in the institutional arrangements to come to a fair and generally supportable decision. If we have the analytical capability will the social capital investments follow or must the higher capacity institutional arrangements come first to articulate the demand for the analytical capability? Why shouldn't it be both together, interactively and neither fast enough, just as it has been in similar chicken and egg problems?

"Now We Can Put a Price on Anything? Do We Want To?" Benefit-cost analysis as developed by civil engineers in the 1930's has long served the water development field well as a tool to say to "No" to those deserving constituents who would like to take on the relatively small share of local match required to have a federal development project. Conflict is now the order of the day for most of these dam and channel projects. The losses and burdens used as a basis for these objections were rarely measurable by the techniques the agencies used with the approval of the Office of Management and Budget. In the last three decades research including that supported by USEPA, and even by the State Water Institute at Cornell that we managed together from 1965 to 1975, has developed techniques to estimate public willingness to pay for almost anything. It has to be described adequately in just a paragraph or two and realistically set in meaningful exchange terms. It's detractors point to the lack of a real market. It's proponents point to the apparently convincing parallels between the survey approach used and the theory that supports the widespread faith in markets as determinants of value. Given the widespread acceptance of the free market as an organizer of society it at least may be a more meaningful indicator of the relative intensity of support for a public choice than the typical opinion survey. Some research indicates that elected officials do positively change their attitudes and behavior when they know what their constituents say they are willing to pay for such services as drinking water protection.

"If Rural and Small are to be Permanent, What Do We Do?" Whether it is the single family houses and tourist services that have made fire fighting such a problem in the forests of the West, or the urge to be rural anywhere in the East, small clusters of water and waste service users and decentralized single on site self providers are here to stay. And note that small community systems violate safe drinking water standards more often that large cities even after adjusting the data for failure to report. The current Source Water Assessment Program should add to awareness of need for change. And perhaps the awareness is growing of the problems on-site waste disposal creates for nearby wells and recreation water bodies. Significant numbers of those TDML targets involve septic system inadequacies. The point, of course, is that the nation can no longer pretend that these are transitional technologies to the eventual conversion to piped central services. Technologies exist to make any site meet any needed standard, and more options appear every year. Instead is the answer to treat local environmental management capacity as the transitional variable? A start would be the capacity to decide which technology to use in place of which ever version of cess pool (eg., Long Island) or tank and tile field is favored. The classic perc test was certainly an improvement in its time, but it still left millions of acres of soils that could not sustain the attack on soil pores by the residuals from almost any onsite treatment system. Even on these soils these are usually installed without even an inspection requirement to further ease the administrative burdens of the local government involved.

For over a hundred years the conventional wisdom to solve "small and rural" has been consolidation of general purpose governments – Woodrow Wilson wrote the definitive book on the subject in 1895. Except, it doesn't happen and the evidence is mounting that it doesn't work very well in the few cases where it has been tried. Is this much like that the conventional wisdom in water resources has been management at the river basin level. Has the nation really tried it? Is it likely to do either by consolidating power in fewer hands? As the Rural Water Associations demonstrate there are many ways to gain the advantages of scale and expertise, and there are many other models to be explored that are more likely to be effective than suggesting broad scale consolidation. Thomas Jefferson may have had it right when he suggested that if the people do not have the insight to make good decisions the answer is not to take that right away from them, but to educate them. We are not sure there is any other choice!

Closing Comment

While we enjoy guessing, we do not know what the next national water policy designs will look like. What we do know is that it is unlikely that water policy at all levels of government will remain unchanged as we enter the next century. Change has emerged, slowly but with certainty, during the last three decades. Stakeholder interests continue to expand. Owners of water rights, private and public, are vigorously pursuing their inherent values. Native American water rights are in a major changing mode. Irrigated agriculture is being questioned, economically and culturally. Endangered species, an understanding of biodiversity and the pursuit of sustainability are changing the landscape. Some speculate that population growth may redefine the mobility rights of Americans. And for sure there is increasing attention is being given to the natural values of water, of waterways, and of their ecosystems
"Our present course is unsustainable – postponing action is no longer an option." With these words, the United Nations Environment Program (UNEP) launched Global Environment Outlook 2000 (GEO.-2000)". Action presupposes defining each opportunity and designing a solution that invites support because those that must support it feel they own it. That means the social capital to follow such steps must be created. Civil engineers have led such investments in the past and we see no reason for them not to do so in the future.

EVOLUTION OF INSTITUTIONAL STRUCTURE FOR WATER RESOURCES PLANNING

Theodore M. Schad[1] and Kyle E. Schilling[2]

INTRODUCTION

In the quarter of a century since the presentation of Schad's 1976[i] paper on the historical development of water resources planning in the United States at the Society's Bicentennial Convention in Philadelphia the nation has experienced a gradual change in water resources planning. Multi-objective and river basin planning with a primary focus on the use of water to promote economic development had peaked in the mid 1970s. Forecasts of lower total demands for water and a shift in concern to environmental quality as a primary objective led to planning based on federal regulaliton for the remainder of the century.

The change has not come about in a linear progression, but in a series of ad hoc steps. The Water Resources Council's (WRC) Principles and Standards for multi-objective planning were replaced by Principles and Guidelines in which economic development was the governing objective, with environmental quality as a secondary, less important objective. The emphasis on federally led planning, with coordinated federal and state efforts by the WRC, also shifted to require increased non-federal cost sharing for planning of traditional water resource programs. At the same time the technological attack on water pollution legislated by the 1972 Water Pollution Control Act Amendments also engendered large expenditures for pollution abatement through regulatory approaches.

As we proceed into the third millennium new emphasis on planning to protect our water infrastructure from terrorist attack and to sustain vital services must also become more prominent in planning for water management. In addition, as overall demand begins to again increase later in the 21st century a return to multi-objective planning including water quality and environmental attributes is needed to effectively use our finite water resources.

BACKGROUND CONDITIONS

The 1976 paper by Schad serves as the basic historical reference to this paper and the reader is referred to it and its included references for more detail on events prior to that time, including the 1973 National Water Commission Report (NWC)[ii]. The author noted in this

[1] Consultant; 4540 25th Road N.; Arlington, VA 22207-4102
[2] Board of Directors, Planning and Management Consultants Ltd; 30 Brooke Crest Lane; Stafford, VA 22554; gschilli@erols.com

paper that there had been a huge proliferation of planning programs in the preceding 15 years, including huge regional studies such as the Northeastern United States Water Supply (NEWS)[iii] study, National Assessments, comprehensive river basin planning under the WRC[iv] and the attempts at regional water quality planning through section 208 studies under the Water Pollution Control Act Amendments of 1972[v]. It was noted that there was a need to better coordinate water quality planning with river basin planning.

In the Carter Administration, 1977 to 1980, a concerted effort was made to evolve a new national water policy for multi-objective water resources management, but it foundered on Congressional concerns that the executive branch was encroaching on legislative branch turf. Subsequently, one of the first actions involving water resources in the Reagan Administration was the elimination of the WRC's planning function and curtailment of the federally supported water resources research programs. As a result, water resources policies tended to be expressions of the goals and aspirations of those constituencies that exercised the greatest control over the law and policy process. Thus federal water policy lurches from crisis to crisis: from ad hoc drought contingency measures during the drought of 1988-1989, to hastily conceived flood relief and protection policies in the aftermath of the 1993 Upper Mississippi River Flood. The Galloway Report[vi], after that flood, noted the increasing costs of floods and relief measures to the nation and the need to adopt a more comprehensive planning framework. The authors note that disaster response and relief costs are almost certainly rising from droughts, as well as floods. The past planning approaches focused on the need for new project development and in addition to not including water quality did not consider flood and drought disaster response and mitigation as part of a full suite of options for water management

In many ways, the mid 1970's were a high point and a turning point in water resources planning. Prior to the 1970's river basin and project planning had almost always been based on projections of ever-increasing demands for water. Large sums of money were being spent on planning in great detail, and insightful comprehensive planning procedures were being developed under the aegis of the WRC. However, only seven of the 18 possible river basin commissions under the WRC authority were ever established. There were also three interstate compact commissions already in existence. The NEWS study was the first major water supply study to provide equal emphasis to demand variables and specific actions to affect future demands; as opposed to focusing solely on development of new supply. Similarly, the 1975-2000 National Assessment, conducted under WRC auspices, in 1979 was the first major comprehensive water resources study to project declines in water use. The WRC forecast proved to be remarkably accurate. According to the United States Geological Survey (USGS), water withdrawals were about 2% less in 1995 than in 1990 and about 10% less than in 1980, the highest use year. However, despite overall decreasing water withdrawals, domestic water supply withdrawals have increased. Significantly, however, there have also been recent decreases in domestic per capita use. In addition, credible projections of water use by the United States Department of Agriculture (USDA)[vii] to 2040 project increased use only in the domestic and public use sectors, as well as for direct livestock consumption.

Water resources and quality issues were also of concern in the latter part of the twentieth century as part of the National awareness of our dependence on developed infrastructure. The National Council on Public Works Improvement (NCPWI)[viii], the first of the Comprehensive "Infrastructure" studies, used a performance perspective to assess infrastructure value emphasizing ability to deliver services and the value of services, as opposed to a simple "needs" approach based on condition and conformance with standards inventories. The final "Fragile Foundations" report card assigned it's highest grade "B" to water resources, compared to a "C" for wastewater; specifically noting the return on investment value of the nation's water resources infrastructure. The NCPWI's Water Resources Report[ix] also recommended, among other things, improved intergovernmental coordinating forums to develop better benefit-cost performance procedures to apply to both proposed new development and existing development, and prophetically with the 1993 floods, a national assessment of levees and flood walls. The NCPWI's Water Resources Report also noted that the over 2000 dams per year being built in the 1960's had declined to a figure in the 1000 's in the 1970's and further still in the 1980's. The "Federal Infrastructure Strategy" (FIS)[x] also validated the performance orientation of "Fragile Foundations" and further developed the principles to adopt outcome-based performance measures. The American Society of Civil Engineers (ASCE) "Infrastructure Report Card" continues to provide a structural condition perspective on all aspects of the Nation's infrastructure and the Critical Water Infrastructure Dialogue[xi], develops a framework for coordination against terrorist threats. Significantly, the events that took place on September 11th, 2001 have added an important reaffirmation of the importance of sustainable water resource systems to deliver not just safe water for consumption but water to maintain economic output, fire-fighting capability and a livable environment.

Recent reports by the Heinz Center[xii] and the Aspen Institute[xiii] have also addressed the need to consider selected dam removal in the suite of river management options, especially, where age, safety and changing needs are concerns. Both reports recognize that orderly planning processes to consider benefits and costs for environmental, social and economic effects need to be developed to fully integrate dam removal into water resources planning, river management and restoration.

POLICY DISCUSSION

Water Resources and institutions to aid decision-making related thereto have been an important part of U.S. history, going back to the development of the Constitution over two hundred years ago. Originally, concerns centered on delineation of the respective rights and responsibilities of the federal and state governments and fostering cooperation among the states. This was followed in the first half of the twentieth century by a gradual increase in federal responsibility and control over water resources which continued into the 1970's. Much of that effort focused on federal studies and actions looking toward development of multi-objective and cooperative federal-state river basin planning. Yet, most major water resources projects had been developed by the mid 1970's and have continued to provide the ability to manage water resources to meet needs. The large scale planning used to plan for water resources development has since been supplanted by a largely federal regulatory system for water quality.

There has been no further attempt to develop a coherent and consistent federal water policy that would guide the actions and decisions of all federal water and related environmental regulatory agencies, including the Corps of Engineers, Environmental Protection Agency, Federal Energy Regulatory Commission, and Fish and Wildlife Service[xiv]. However, the crises requiring ad hoc drought relief and contingency measures during the drought of 1988-89, and hastily conceived flood relief and protection policies in the aftermath of 1993 upper Mississippi River flood demonstrate continuing need for thoughtful preemptory policy development and integration with disaster response and mitigation policies, as well.

The federaly funded WRC and the many river basin commissions set up to assist in the development oriented planning processes no longer exist[xv]. The last quarter of the 20th century was dominated by a transition from traditional water quantity development approaches to water quality and environmental issues, largely because basic quantity needs had been met[xvi]. There are successes: freshwater withdrawals have decreased steadily from their high in 1980; existing planned infrastructure has met changing demands; water pollution is decreasing; and aquatic ecosystems are rebounding. Yet; without a coherent federal water policy, the less than efficient amalgamation of regulatory policies will increasingly hamper the effectiveness of prospects for innovations in watershed and ecosystem management. Federal environmental policy is being coordinated centrally through the Council on Environmental Quality (CEQ), which is not equipped to deal with water policy for broader water resources issues such as floods and droughts. The ad-hoc and episodic response to recent severe floods and droughts provides ample evidence.

A panel convened by the National Academy of Public Administration (NAPA)[xvii] in response to a congressional requirement also stated flatly that "The nation's current environmental protection system cannot deliver the healthy and sustaining future that Americans want." Thus suggesting that EPA will have "to use 21st century tools to address the problems of the 21st century. Clearly even though the nation's most basic quality and quantity needs are being met, decision-making for water management in the U.S. is becoming increasing complex. In addition, public awareness of water resources is increasing. The 1996-1997 River and Conservation Directory lists more than 3,000 citizen watershed-related organizations in the U.S. TheEPA has established and funded various financial and educational assistance programs to assist this process. In addition, there are now numerous watershed conferences and much watershed rhetoric in the U.S., as well as activities like the American Heritage Rivers program.

An apparent anomaly to the trend against river basin organizations is the effort by three states -Alabama, Florida and Georgia - to form new interstate compacts for the Appalachicola, Chattahoochee and Flint (ACF) Rivers and for Alabama, Coosa and Tallapoosa (ACT) Rivers. This new effort to establish forums to facilitate allocation of water between the states in very contentious settings appears to be at least partially based on the application of shared vision planning models. The Federal government, although a party to the compacts, is subject to a veto by any of the three states for any compact decision. Although detailed benefit-cost and multi-objective planning procedures are not being used in the emerging shared vision planning processes; they do indicate increasing awareness of the need to compare and plan for alternative courses of action including water quality along the lines suggested by the NAPA panel.

The recently held Critical Water Infrastructure Dialogue has, in addition, reaffirmed a basic tenet, going all the way back to constitutional discussions, of U.S. water resources development for sustainable, secure delivery of water to provide, economic and life support services. The dialogue participants also recognized the increased need to value safety and the environment in the provision of these services.

CONCLUSION

The authors present this hastily prepared paper in the hope that its fragmentary summary of what has been happening in the field of water resources management in the last quarter century will start a new dialogue that will lead to institutional changes to improve water management in the future. They note that others have contributed valuable insights at the end of the twentieth century to advance thinking and dialogue, as well.[xviii] Recent studies by NAPA panels indicate that the environmental protection programs authorized by the Congress are not being conducted by the EPA in ways that will achieve the desired objectives. The planning of the water pollution abatement program and its coordination with the programs of other water resources agencies authorized by Sections 208 and 209 of the 1972 Water Pollution Control Act have also never been carried out. Recognition of changing conditions and limited knowledge to predict ecosystem response to change was recognized by the Water Science and Technology Board (WSTB)[xix] of the National Research Council as a basic reason to recommend periodic review for and adaptive management of water resource systems. It is also considered necessary to use improved flexible benefit-cost analysis when considering dam removal for safety and environmental restoration in a process to create balance and compatible techniques to assess water resources system changes including existing development. Panels created by the WSTB have also indicated areas to improve the planning process used by the U. S. Army Corps of Engineers for both project justification and management of existing projects. Yet, the valuable contributions to the nation's water resources infrastructure, as recognized in "Fragile Foundations"; maintenance needs as documented by the ASCE Report Card and the need to maintain delivery of vital water services in the face of terrorism all increase the importance of improved water quality and quantity planning in the 21st century.   Consequently, the authors recommend a renewed commitment to increase federal leadership to integrate water quality, quantity, disaster management and environmental planning and coordination in a cooperative local, state, federal manner.

The authors believe that past national policy and study efforts have improved understanding of the issues and incrementally resulted in adaptive management behavior to improve the development and management of the nation's water resources. They also believe that such efforts including coordination through Councils such as, WRC or CEQ are needed to maintain and continue the progress that we have made.

REFERENCES

[i] "Water Resources Planning – Historical Development", Theodore M. Schad, J. Water Resources Planning and Management Division, 105, 9.-25

[ii] "Water Policies for the Future", United States National Water Commission, United States Government Printing Office, Washington, D.C., June 14, 1973

[iii] Northeastern US Water Supply Study, Title I, PL 89-298

[iv] Water Resources Planning Act, PL 89-80, July 22, 1965

[v] Federal Water Pollution Act Amendments, PL 92-500, October 18, 1972

[vi] Sharing the Challenge: Floodplain Managtement into the 21st Century (The Galloway Report) U.S. Interagency Floodplain Management Review Committee, U.S.G.P.O. Washington, D.C. 1994

[vii] "Past and Future Freshwater use in the United States", U.S. Department of Agriculture, Forest Service, 1990

[viii] "Fragile Foundations: A Report on American's Public Works", National Council on Public Works Improvement, February, 1988.

[ix] "The Nation's Public Works: Report on Water Resources", National Council on Public Works Improvement, May 1987.

[x] "Living within constraints: An emerging Vision for High Performance Public Works" U.S. Army Corps of Engineers, Institute for Water Resources, January 1995.

[xi] "Critical Water Infrastructure: A Beginning Dialogue – Networking U.S. Water Assocations-Organizations, and Agencies", Environmental and Water Resources Institute of the American Society of Civil Engineers, 2002

[xii] "Dam Removal Science and Decision Making", The Heinz Center, 2002

[xiii] "Dialogue on Dams and Rivers" The Aspen Institute 2002

[xiv] "Water Policy: Who Should Do What", Theodore M. Schad, Universities Council on Water Resources Update, No. 111 Spring 1998.

[xv] "Reflections on Water Management, Academia and Disciplinary Perspectives in the U.S., Kyle E. Schilling, Universities Council on Water Resources Update No 123, June 2002

[xvi] "The Future for Water Resources Planning and Decision-making Models", Kyle E. Schilling Universities Council on Water Resources Update, No 111, Spring 1998

[xvii] "Transforming Environmental Protection for the 21st Century", National Academy of Public Administration 2002

[xviii] "Reflections on a Century of Water Science and Policy", Universities Council on Water Resources, update, No.116, March 2000, Charles W. Howe, Editor.

[xix] "Envisioning the Agenda for Water Resources Research in the Twenty First Century", National Research Council, Water Science and Technology Board, 2001

HISTORY OF ASCE'S HYDROLOGY HANDBOOK

Kenneth G. Renard[1]

In 1949, the American Society of Civil Engineers (ASCE) prepared a hydrology handbook titled *Manual of Engineering Practice No. 28* and published by a committee of the Hydraulics Division. It was subsequently reprinted in 1952 and 1957. Prior to another printing in 1982, it was decided that a new Manual be prepared. With encouragement from the Executive Committees of ASCE's Hydraulic Division and the Irrigation and Drainage Division, a Task Committee was identified to prepare a new *Manual*. The Task Committee subsequently prepared a table of contents which served as a guide for a new handbook under the control ASCE's Management Group D, the Group that oversaw the activities of five ASCE technical divisions involved in *water engineering*. Under the leadership of Conrad G. Keyes, Jr., Management Group D established a committee with two members from the Hydraulics Division and two from the Irrigation and Drainage Division. The four member committee (Thomas P. Wootton, Chair, Catalino B. Cecilio, Lloyd C. Fowler, and Samuel L. Hui) subsequently directed the preparation of the new **Hydrology Handbook.** Completion of the book took from its beginning in 1988 until its completion in 1995.

The outline of the new book and the chapter authors follows:

Chapter 1: Introduction
Lloyd C. Fowler

Chapter 2: Precipitation
Clayton L. Hanson, Chair: Karl L. Gebhardt; Gregory L. Johnson; Marshal J. McFarland; and James A. Smith

Chapter 3: Infiltration
Walter J. Rawls, Chair: David Goldman; Joseph A. Van Mullen; and Timothy J. Ward

Chapter 4: Evaporation and Transpiration
Richard G. Allen and William O. Pruitt, Co-chairs: Jost A. Businger; Leo J. Fritschen; Marvin E. Jensen; and Frank H. Quinn

Chapter 5: Ground Water
Lloyd C. Fowler, Chair: Nazer Ahmed; Keith E. Anderson;Thomas P. Ballestero; Ronald K. Blatchley; Carl H. Carpenter; Clyde S. Conover; Marvin V. Damm; Donald J. Finlayson; James D. Goff; A. Ivan Johnson; W. Martin Roche

---

[1] USDA-ARS-SWRC, 2000 E. Allen Road, Tucson, AZ 85719-1596; Krenard@Tucson.ars.ag.gov

Chapter 6: Runoff, Streamflow, Reservoir Yield, and Water Quality

Anand Prakash, Chair: Richard J. Heggen; Victor M. Ponce; John A. Replogle; and Henry C. Riggs

Chapter 7: Snow and Snowmelt

Douglas D. Speers, Chair: George D. Ashton; and David M. Rockwood

Chapter 8: Floods

Bi-Huei Wang, Chair: Michael L. Anderson; Gary R. Dyhouse; Vernon K. Hagen; Khalid Jawed; John R. Riedel; and Jery R. Stedinger

Chapter 9: Urban Hydrology

David F. Kibler, Chair: A. Osman Akan; Gert Aron; Christopher B. Burke; Mark W. Glidden; and Richard H. McCuen

Chapter 10: Water Waves

Zeki Demirbilek, Chair: Robert A. Dalrymple; Robert M. Sorenson; Edward F. Thompson; J. Richard Weggel

Chapter 11: Hydrologic Study Formulation and Assessment

John C. Peters, Chair: Shawna Anderson; Patrick Atchison; John J. Buckley; David Ford, Katherine Hon; and Richard H. McCuen

Numerous people served as reviewers of individual chapters and contributed extensively to the chapter completeness. As the chapters were finalized, Richard K. Heggen did the overall editing.

Finally, a blue ribbon group served as reviewers of the entire book or individual chapters. Thus the entire new **Hydrology Handbook** represents the effort of in excess of 100 individuals. The 784 page book serves as a baseline document for the engineering and hydrology profession.

Watershed Models

Vijay P. Singh[1] and Donald K. Frevert[2]

Abstract: Before the 1960s, watershed modeling was primarily confined to modeling of individual components of the hydrologic cycle. Since the advent of computers and the ground breaking development of the Stanford watershed model, numerous watershed models have been developed. This study traces the history of watershed modeling.

Introduction

Before the 1960s, watershed modeling was primarily confined to modeling of individual components of the hydrologic cycle, principally due to limited computing capability and limited data. Surface runoff modeling using the unit hydrograph theory and the determination of peak runoff using the rational method received much emphasis. Infiltration models were developed for agricultural soils and remained relatively simple for hydrologic applications. Evapotranspiration models, based on radiation, temperature, mass transfer, energy balance or combinations thereof, were developed for agricultural water management. For flood forecasting, simple flood wave propagation models, such as the Muskingum method, were developed. The SCS-CN method was and still is popular for computing the amount of runoff from agricultural and forest and range lands. The emphasis was on simple analytical methods. Since the advent of computers in the 1960s and the ground breaking development of the Stanford watershed model by Crawford and Linsley (1966), numerous watershed models have been developed. The development of more integrated and comprehensive models continues today. This study traces the history of watershed models before and during the computer era.

Component Models

The beginnings of hydrological modeling can be traced to the development of civil engineering in the nineteenth century for design of roads, canals, city sewers, drainage systems, dams, culverts, bridges, and water supply systems. Until the middle of the 1960s, hydrologic modeling involved development of concepts, theories and models of individual components of the hydrologic cycle relating to overland flow, channel flow, infiltration, depression storage, evaporation, interception, and base flow. For runoff generation, the Hortonian mechanism,

[1] Arthur K. Barton Prof., Dept. of Civil & Env. Engrg., Louisiana St. Univ., Baton Rouge, LA 70803-6405, (e-mail: cesing@lsu.edu)

[2] Hydr. Engineer, River Systems and Met. Group, USBR, Tech. Service Center, Denver Federal Center, P. O. Box 25007, Denver, CO 80225-0007

subsurface flow mechanism and partial and source area contribution were developed. In what follows, only a short discussion of the development of component models is given.

Areal Precipitation

Precipitation constitutes input to watershed hydrology models. In many models, an areal estimate of precipitation is made. Mean areal precipitation has been estimated using a number of techniques but the arithmetic mean, Thiessen polygons and isohyetal methods are still popular. Trend surface analysis has also been popular. Lately, the focus has shifted toward using spatially measured precipitation, as for example, by segmenting a watershed into hydrologic response units or subbasins, with each subbasin being represented by a raingage.

Watershed Representation

Before the computer era, a watershed was represented as a black box and no consideration was given to its spatial variability. Today a watershed is represented by (1) grid methods or (2) conceptual methods (Singh, 1996). A grid method attempts to maintain model flow patterns similar to those in the prototype watershed response. Conceptual methods represent watershed geometry using a network of elemental sections which may be fictitious such as reservoir and channel or real such as plane, triangular section, converging section, diverging section, and channel. Depending on the arrangement of these elements, many simplified geometric configurations have been employed, including V-shaped geometry, composite geometry, cascade of planes and channels, and complex configurations of planes and channels. If a watershed is represented as a distributed system, then its subunit delineation may be on the basis of geomorphologic, conceptual, digital terrain, digital elevation, segmentation, or hydrologic response unit considerations.

Surface Runoff

The origin of runoff modeling dates back to the rational method developed by Mulvany in 1850 for relating storm runoff peak to rainfall intensity. Sherman (1932) introduced the unit hydrograph (UH) concept for relating the direct runoff response to rainfall excess. The UH concept has since been investigated extensively along three lines: (1) derivation of the UH, (2) synthesis of the UH for ungaged watersheds, and (3) development of the UH theory and conceptual models. Using watershed characteristics, Snyder (1938) and Taylor and Scharwz (1952) developed methods for synthesizing the UH. The Soil Conservation Service (1955) proposed an average dimensionless UH which was later approximated as a triangle

Employing the theory of linear systems, Nash (1957) and Dooge (1959) developed the theory of the instantaneous unit hydrograph (IUH). Using geomorphologic properties, Rodriguez-Iturbe and Valdes (1979) developed a probabilistic approach for derivation of the IUH by explicitly incorporating the drainage basin laws.

In 1955, Lighthill and Whitham developed the kinematic wave theory for flow routing in long rivers. Woolhiser and Liggett (1967) have shown that the kinematic wave approximation is sufficiently accurate for modeling overland flow. Morris and Woolhiser (1980) introduced the diffusion wave approximation to the St. Venant equations of overland flow.

Infiltration

The earliest attempts for computing rain infiltration were the phi-index, W-index, and the equations of Kostiakov (1932), and Horton (1939). Using physical principles, Philip (1957), and Smith (1972), among others, derived infiltration models with physical parameters. Employing the systems approach, Singh and Yu (1990) derived a generalized infiltration model which specializes into many popular models mentioned above.

Subsurface Flow and Interflow

Runoff is produced by one or a combination of three processes: (1) the infiltration excess (Horton) mechanism, (2) the saturation (Dunne) mechanism, and (3) the subsurface mechanism. In arid and semi-arid regions, the Horton mechanism is the dominant mechanism. In humid and subhumid regions, the saturation mechanism is dominant. This mechanism has led to the concept of partial area or source area contribution. The subsurface mechanism is commonly found in forest watersheds. Many watershed hydrology models employ the Horton mechanism even though it may not necessarily be the one producing runoff. Some models account for all of these mechanisms. Dunne (1978) summarized hillslope process studies.

Hewlett (1961) has shown that unsaturated flow could generate streamflow, and this has given rise to the subsurface flow mechanism. In the 1970s the variable source area or partial source concept received significant attention in the context of vegetated or forested lands. The variable source area incorporates the entire range of hillslope processes. Interflow occurs as near-surface flow of water within the soil profile resulting in seepage to a stream channel, and may involve both saturated and unsaturated flows in vertical and downslope directions. This may encompass all subsurface contributions to the storm hydrograph. Henderson and Wooding (1964) applied the kinematic wave theory to model both buildup and decay of subsurface flow. Beven (1981) compared the Dupuit-Forchheimer (DF) approximation, the extended DF approximation and the kinematic wave theory for modeling subsurface flow.

Groundwater Flow and Baseflow

Combining Darcy's law with the continuity equation, Theis (1935) derived the relation between the lowering of the piezometric surface and the rate and duration of well discharge using ground-water storage, which laid the foundation for quantitative groundwater hydrology. The study of groundwater and infiltration led to development of techniques for separation of baseflow and interflow in a hydrograph (Barnes,

1940). Although the groundwater flow theory is quite well developed, baseflow is still commonly modeled using lumped empirical methods which employ linear reservoirs (Dooge, 1960).

Evaporation and Evapotranspiration

Models for computing evaporation from free water surfaces have been based on water budget, mass transfer, energy balance, and combinations of energy balance and mass transfer. Penman (1948) combined the energy balance and mass transfer methods leading to a pioneering theory of evaporation. This theory has since been the catalyst for most later developments in the area of evaporation and evapotranspiration. Monteith (1965) incorporated aerodynamic and canopy resistance in the Penman equation and the modified equation is popularly known as the Penman-Monteith equation. The various variants of the Penman-Monteith method have been summarized by Allen (1985).

Evaporation from vegetated soils gives rise to evapotranspiration (ET) which depends on the soil moisture and rootzone development. Blaney and Criddle (1945) developed an empirical relation between ET, mean air temperature and mean percentage of daytime hours. This has been one of the most popular models for computation of ET. Thornthwaite (1948) developed a popular equation for computing (ET) under limited water conditions. Using solar radiation and temperature, Jensen and Haise (1963) developed another popular ET model. Each of these models has been employed in watershed hydrology modeling. Generally the rate of actual evapotranspiration is needed, but its computation is complicated by varying soil moisture, especially when it is below field capacity, and the complicated nature of root zone development.

Abstraction

Interception, depression storage, and detention storage constitute abstractions which have received relatively little attention. Interception has been treated well but mostly in forested watersheds. Horton (1919) derived a series of empirical formulas for estimating interception during a storm for various types of vegetal covers. Detention and depression storage is modeled using an exponential model. For computing the runoff volume of a storm considering abstractions, the Soil Conservation Service (SCS) of the U. S. Department of Agriculture developed in 1956 what is now referred to as the SCS-CN method. The method has also been extended to model infiltration, runoff hydrographs, water quality modeling and continuous hydrologic simulation.

Rainfall-Excess/Soil Moisture Accounting

Once all abstractions are determined, a soil moisture accounting is done, which requires an application of the water budget equation. If the portion of precipitation infiltrating into the soil is estimated, the status of soil moisture is assessed, which is needed to determine the watershed wetness and transform potential evapotranspiration (PE) into actual evapotranspiration (AE). The computation of actual evapotranspiration has been done in different ways, but in all cases the methods have been empirical and a sound theory is generally lacking.

Snowmelt Runoff

Although snow ablation and snow melting processes are understood reasonably well, most models estimate snowmelt runoff empirically using a degree-day method or an energy-based method. U. S. Army Corps of Engineers (1956) developed a degree-day method for computing snowmelt and also derived regression equations. For modeling snowmelt runoff, Colbeck (1974) developed a kinematic wave theory. Snowmelt runoff routing is also done empirically or using the kinematic wave method. Sometimes, a unit hydrograph approach is also employed. Singh and Singh (2001) presented a discussion of many of the snowmelt models.

Stream-Aquifer Interaction

Stream-aquifer interaction models are well developed independently but watershed models seem to have taken limited advantage of their development. The interaction models are based on systems theory or groundwater hydraulics. The systems theory is based on linear reservoirs or linearized parabolic partial differential equations. Bathala et al. (1980), among others, modeled stream-well-aquifer interactions using linearized parabolic differential equations.

Reservoir Flow Routing

In 1928 Puls of the U. S. Army Corps of Engineers, Chattanooga District, developed a method for flow routing through reservoirs, assuming invariable storage-discharge relationships and neglecting the variable slope during flood propagation. This method, later modified by the U. S. Bureau of Reclamation (1949), is now referred to as the modified Puls method or the storage indication method. For flood routing through reservoirs, this is still one of the most popular methods in hydrology. Currently, diffusion wave theory and dynamic wave theories are also being employed in watershed models (Fread, 1984).

Channel Flow Routing

Using the concept of wedge and prism storage, McCarthy and others developed the Muskingum method of flow routing in connection with the Muskingum Conservancy District Flood Control Project in 1934-35. The method was based on the spatially lumped form of continuity equation and a volumetric flux relation expressing storage as a linear function of weighted inflow and outflow discharges. This method has since been a popular method of flood routing.

Flow routing methods based on the St. Venant equations and their simplifications through kinematic wave and diffusion wave approximations were also developed. A number of linear dynamic wave models were developed (Singh, 1996). Cunge (1969) showed the link between the Muskingum method and the convection-diffusion equation. Using a semi-infinite channel, Hayami (1951) derived an analytical solution for a linear diffusion wave equation. Beginning with the work of Lighthill and Witham (1955), kinematic wave routing methods were developed.

Water Quality

The 1980s and the 1990s also witnessed the linking of hydrologic models with those of geochemistry, and environmental biology for two reasons. First, there was increased understanding of spatial variability of hydrologic processes and the role of scaling. This was essential because different processes operate at different scales and linking them to develop an integrated model is always challenging. Second, the digital revolution made possible the employment of GIS, remote sensing techniques and data base management systems. Currently, a number of watershed hydrology models have water quality components.

Model Calibration

Prior to the 1960s, parameters of component models were determined based on graphical fit or least squares. The least squares and the Nash-Sutcliffe efficiency objective functions were frequently employed. Diskin and Simon (1977) proposed guidelines for selecting an objective function in model calibration. Several optimization methods, including direct search methods, gradient search methods, random search methods, multi-start algorithms, and shuffled complex algorithms, have been developed. To facilitate finding a unique "best" parameter set (and the inadequacy of optimization methods) for multi-input-ouput hydrologic models, generalized likelihood uncertainty estimation (GLUE), the Monte Carlo membership set procedure (MCSM), and the prediction uncertainty method (PU) were proposed. Application of artificial neural networks (ANNs) to hydrologic modeling has lately been receiving considerable attention. Govindaraju and Rao (2000) contains a variety of applications of ANNs and genetic algorithms to hydrologic modeling.

Model Testing

Watershed models are verified using a split-sample approach, Monte Carlo simulation, assessment of model uncertainty, and propagation of errors. Recognizing the model uncertainty and critically analyzing the limitations of the existing model testing methods, Kuczera and Franks (2002) developed a probabilistic framework for model testing, including a Bayesian paradigm, articulation of errors, and data augmentation strategies. What is usually not done is to assess the validity of the model for a range of conditions or for a variety of data sets and to delineate limitations for which the model is valid.

Stochastic Models

Another popular approach to total watershed modeling is stochastic hydrology. Yevjevich (1972) did much of the pioneering work in this field. This approach preserves the statistical properties of historically observed watershed runoff time series and uses those properties as the basis to stochastically generate hydrologic traces which might be encountered in the future. Much of the detailed theory which underlies time series modeling in general and stochastic hydrology in particular is described in Salas, et al (1980).

In the late 1970's, Lane developed a generally applicable stochastic hydrology computer package known as Lane's Applied Stochastic Techniques or LAST. LAST, in it's most recent form, is described in Lane and Frevert (1990). During the early 1980's, Grygier and Stedinger (1990) developed the Synthetic Streamflow Generation Software Package, commonly known as SPIGOT.

During the 1990s, Salas and other researchers at Colorado State University recognized the need to update and modernize the capabilities which were available in LAST. The culmination of this effort was the Stochastic Analysis, Modeling and Simulation Package (SAMS). A detailed description of SAMS is provided in Singh and Frevert (2002b).

Current Trends

Advances in watershed modeling have occurred at an unprecedented pace since the development of Stanford Watershed Model (SWM or now HSPF), with emphasis on physically based models. During the decades of the 1970s and the 1980s, a number of mathematical models were developed not only for simulation of watershed hydrology but also for their applications in other areas, such as environmental and ecosystems management. The development of new models or improvement of the previously developed models has continued till today. This has been possible partly due to developments in data acquisition through remote sensing and satellite technology, chemical tracers, and digital elevation and terrains models; data processing through geographical information systems; database management systems; graphical displays; and growing sophistication in computing software and computing prowess. Nowadays many federal agencies in the United States have their own models or some variants of models developed elsewhere. In 1991, the U. S. Bureau of Reclamation prepared an inventory of 64 hydrology and water resources models classified into 4 categories and the inventory is currently being updated. Burton (1993) compiled Proceedings of the Federal Interagency Workshop on Hydrologic Modeling Demands for the 1990s, which contain several important watershed hydrology models. Singh (1995) edited a book that summarized 26 popular models from around the globe. The Subcommittee on Hydrology of the Interagency Advisory Committee on Water Data (1998) published proceedings of the First Federal Interagency Hydrologic Modeling Conference which contains many popular watershed hydrology models developed by federal agencies in the United States. Wurbs (1998) listed a number of generalized water resources simulation models in seven categories and discussed their dissemination. Singh and Frevert (2002 a, b) edited two books that contain 38 models.

There are several models that are well-known general watershed models in current use in the U. S. and elsewhere. Examples of such watershed hydrology models are SWMM, PRMS, NWS River Forecast System, SSARR, Systme Hydrologique

Europeen (SHE), TOPMODEL, and so on. All of these models have since been significantly improved. SWM, now called HSPF, is far more comprehensive than its original version. For example, SHE has been extended to include sediment transport and is applicable at the scale of a river basin. TOPMODEL has been extended to contain increased catchment information, more physically based processes and improved parameter estimation.

These models vary significantly in the model construct of each individual component process partly because these models serve somewhat different purposes. HEC-HMS is considered the standard model in the private sector in the U. S. for design of drainage systems, quantifying the effect of land use change on flooding, etc. The NWS model is the standard model for flood forecasting. HSPF and its extended water quality model are the standard models adopted by the Environmental Protection Agency for EPA-related work. The MMS model of the U. S. G. S. is a widely used model for water resources planning and management works, including some of those under the purview of the U. S. Bureau of Reclamation.

The 1980s and the 1990s also witnessed the linking of hydrologic process models with those of reservoir management. One prominent example of linking between models is the Watershed and River Systems Management Program which is supported by the US Bureau of Reclamation, the US Geological Survey and a number of other partners. This program links RiverWare – a generally applicable reservoir management framework with MMS, an hydrologic data base known as HDB and the SAMS modeling package. Plans are underway to facilitate the use of these tools in conjunction with other specialized models including agricultural water demand, fisheries, economics, water quality and ground water. Detailed descriptions of the products of this program are provided in Frevert, et al (2000).

CONCLUSIONS

Many current watershed models are comprehensive, distributed and physically based. They possess the capability to accurately simulate watershed hydrology and can be applied to address a wide range of environmental and water resources problems. Some of these models are also capable of simulating water quality. The models are becoming embedded in modeling systems whose mission is much larger, encompassing several disciplinary areas.

REFERENCES

Allen, R. G. (1985). "A Penman formula for all seasons." *J. Irrig. & Drain. Engrg.*, ASCE, 112(4), 348-368.

Barnes, B. S. (1940). "Discussion on analysis of runoff characteristics by O. H. Meyer." *Trans., Am. Soc. Civil Engineers*, 105, 104-106.

Bathala, C. T., Rao, A. R. and Spooner, J. A. (1980). "Linear system models for regional aquifer evaluation studies." *Water Resourc. Res.*, 16(2), 409-422.

Beven, K. J. (1981). "Kinematic subsurface stormflow." *Water Resourc. Res.*, 17(5), 1419-1424.

Blaney, H. F. and Criddle, W. D. (1945). "Determining water requirements in irrigated areas from climatological data." *Report No. 17*, Soil Cons. Service, U. S. Dept. Agric., Washington, D. C.

Bureau of Reclamation (1991). "Inventory of hydrologic models." *Report,* U. S. Depart. Interior, Denver, Colorado.

Burton, J. S. compiler (1993). « Proceedings of the Federal Interagency Workshop on Hydrologic Modeling Demands for the 90's." *U. S. Geolog. Survey Water Resourc. Invest. Rep.t* 93-4018, Federal Center, Denver, Colorado.

Colbeck, S.C. (1974). Water flow through snow overlying an impermeable boundary. *Water Resourc. Res.*, 10, 119-123.

Crawford, N. and Linsley, R. K. (1966). "*Digital simulation in hydrology: Stanford watershed model*." Tech. Rept. No.39, Stanford Univ., Palo Alto, Cal.

Cunge, J. (1969). "On the subject of a flood propagation computation method (Muskingum method)." *J. Hydraul. Res.*, 7(2), 205-230.

Diskin, M. H. and Simon, E. (1977). "A procedure for the selection of objective functions for hydrologic simulation models." *J. Hydrol.*, 34, 129-149.

Dooge, J. C. I. 1959. "A general theory of the unit hydrograph." *J. Geophys. Res.*, 64(2), 241-256.

Dooge, J. C. I. (1960). "The routing of groundwater recharge through typical elements of linear storage." *IASH Pub.* No. 52, 286-300.

Dunne, T. (1978). "Field studies of hillslope flow processes." in *Hillslope Hydrology*, edited by M. J. Kirkby, Wiley Interscience, 227-293.

Fread, D.L. (1984). "Flood routing." In *Hydrological Forecasting*, edited M. G. Anderson and T. P. Burt, Chapt. 14, Wiley, New York.

Frevert, D., Lins, H., Fulp, T., Leavesley, G. and Zagona, E. (2000). "The Watershed and River Systems Management Program – An Overview of Capabilities." *Proc. Watershed and Operations Management 2000 Conf.*, ASCE, Reston, Virginia.

Govindaraju, R. S. and Rao, A. R., editors, (2000). "*Artificial Neural Networks in Hydrology.*" 329 pp., Kluwer Acad. Publishers, Boston.

Grygier, J.C. and Stedinger, J.R. (1990). "SPIGOT, A Synthetic Streamflow Generation Software Package." *Tech. Descript. Version 2.5*, School of Civil and Env. Engrg., Cornell Univ., Ithaca, New York.

Hayami, S. (1951). "On the propagation of flood waves." *Bulletin 1*, Disaster Prevention Res. Inst., Kyoto, Japan.

Henderson, F. M. and Wooding, R. A. (1964). "Overland flow and groundwater flow from a steady rainfall of finite duration." *J. Geophys. Res.*, 69(8), 1531-1540.

Hewlett, J. D. (1961). "Soil moisture as a source of base flow from steep mountain watersheds." U. S. D. A.-FS, Southeast For. Exp. St., *Paper No. 132*.

Horton, R. E. (1919). "Rainfall interception." *Month. Weather Rev.*, 147, 603-623.

Horton, R. E. (1939). "Analysis of runoff plot experiments with varying infiltration capacities." *Trans., Am. Geophys. Union*, 20(IV), 683-694.

Jensen, M. E. and Haise, H. R. (1963). "Estimating evapotranspiration from solar radiation." *J. Irrig. & Drain. Div.*, ASCE, 89, 15-41.

Kostiakov, A. M. (1932). "On the dynamics of the coefficient of water percolation in soils and of the necessity of studying it from a dynamic point of view for purposes of amelioration." *Trans., 6th Comm.*, Int. Soil Sci. Soc., Russian, Part 1, pp. 17-29.

Kuczera, G. and Franks, S. W. (2002). "Testing hydrological models: fortification or falsification." Chapt. 5 in *Mathematical Models of Large watershed Hydrology*, edited by V. P. Singh and D. K. Frevert, Water Resourc. Publ., Litleton, Co.

Lane, W.L. and Frevert, D.K. (1990). "Applied Stochastic Techniques." *Personal Computer Version 5.2*, Users Manual, U. S. Bureau of Reclam., Denver, Co.

Lighthill, M. J. and Whitham, G. B. (1955). "On kinematic waves:1. Flood movement in long rivers." *Proceedings, Royal Society*, London, Series A, 229, 281-316.

Monteith, J. L. (1965). "Evaporation and environment." *Symp., Soc. Exp. Bio.*, 19, 205-235.

Morris, E. M. and Woolhiser, D. A. (1980). "Unsteady one dimensional flow over a plane: partial equilibrium and recession hydrographs." *Water Resourc. Res.*, 16(2), 355-360.

Nash, J. E. (1957). "The form of the instantaneous unit hydrograph." *Hydrological Sciences Bulletin*, 3, 114-121.

Penman, H. L. (1948). "Natural evaporation from open water, bare soil and grass." *Proc. Royal Soc. (London), Ser. A*. 193, 120-145.

Philip, J. R. (1957). "Theory of infiltration: sorptivity and algebraic equations." *Soil Science*, 84, 257-265.

Rodriguez-Iturbe, I. and Valdes, J. B. (1979). "The geomorphologic structure of hydrologic response." Water Resourc. Res., 15(6), 1409-1420.

Salas, J.D., Delleur, J.W., Yevjevich, V. and Lane, W.L. (1980). *"Applied Modeling of Hydrologic Time Series."* Water Resourc. Pubs., Littleton, Co.

Sherman, L. K. (1932). "Stream flow from rainfall by the unit graph method." *Engineerg. News Rec..* 108, 501-505.

Singh, V. P., editor. (1995). *"Computer models of watershed hydrology."* Water Resourc. Pubs., Littleton, Co.

Singh, V. P. (1996). *"Kinematic wave modeling in water resources: surface water hydrology."* John Wiley & Sons, New York.

Singh, V. P. and Frevert, D. K., editors (2002a). *"Mathematical Models of Large Watershed Hydrology."* Water Resourc. Pub, Highlands Ranch, Co.

Singh, V. P. and Frevert, D. K., editors (2002b). *Mathematical Models of Small Watershed Hydrology and Applications.* Water Resourc.Pubs., Highlands Ranch, Co.

Singh, P. and Singh, V. P. (2001). *"Snow and Glacier Hydrology."* Kluwer Acad. Pubs., Dordrecht, The Netherlands.

Singh, V. P. and Yu, F. X. (1990). "Derivation of an infiltration equation using a systems approach." *J. Irrig. & Drain. Engrg.*, ASCE, 116 (6), 837-858.

Smith, R. E. (1972). "The infiltration envelope: results from a theoretical infiltrometer." J. Hydro., 17, 1-21.

Snyder, F. F. (1938). "Synthetic unit graphs." Trans. Am. Geophys. Uni., 19, 447-454.

Soil Conservation Service (1955). "Hydrology." *SCS National Engrg.Handbook*, Section 4, U. S. Dept. Agric., Washington, D. C.

Subcommittee on Hydrology (1998). "Proceedings of the First Federal Interagency Hydrologic Modeling Conference." Interagency Adv. Comm. Water, U. S. Geolog. Surv., Reston, VA.

Taylor, A. B. and Scharwz, H. E. (1952). "Unit hydrograph lag and peak flow related to basin characteristics." *Trans. Am. Geophys. Uni.*, 33, 235-246.

Theis, C. V. (1935). "The relation between the lowering of the piezometric surface and the rate and duration of discharge of a well using ground-water storage." *Trans., Am. Geophys. Uni.*, 16, 519-524.

Thornthwaite, C. W. (1948). "An approach toward a rational classification of climate." *Geograh. Rev.*, 38, 55-94.

U. S. Army Corps of Engineers (1956). "Snow Hydrology-Summary Report of Snow Investigations." North Pac. Div., Dept. Army, Portland, Oregon.

U. S. Bureau of Reclamation (1949). "Flood routing." Chapt. 6.10 in *Flood Hydrology*, Pt. 6, in Water Studies, Vol. IV, Washington, D. C.

Woolhiser, D. A. and Liggett, J. A. (1967). "Unsteady one dimensional flow over a plane: the rising hydrograph." *Water Resourc. Res.*, 3(3), 753-771.

Wurbs, R. A. (1998). "Dissemination of generalized water resources models in the United States." *Water International.* 23, 190-198.

Yevjevich, V. (1972)."Stochastic Processes in Hydrology." *Water Resourc. Pubs.*, Littleton, Co.

Revisiting *Rivers of History:*
Another Look at Life on the Coosa, Tallapoosa, Cahaba, and Alabama Rivers

Harvey H. Jackson III*

It was a coincidence. I had just received the invitation to submit this article to the EWRI *Proceedings* when the "Coosa-Alabama River Improvement Association Newsletter" arrived to announce that once again President Bush's budget did not include funding to maintain and operate a navigation channel on the Alabama River. Why? Because by federal criteria used to assess navigation projects this stream, which was once crowded with all sorts of watercraft, is now an "unproductive" river.[1]

That got me thinking -- thinking about what I had written back in the early 1990s when I was pulling together the loose ends that, when tied up, would be *Rivers of History: Life on the Coosa, Tallapoosa, Cahaba, and Alabama.* And thinking about what I would say now, to explain why this once important river system, today, isn't. To do this, to make sense of the rise and decline of this remarkable waterway, I need to go back, back to where I began the book, back to the rivers before man came to change them, back to when they still ran wild and free.

First there was the Coosa, or rather the Coosas; for once there were two rivers. The first was born where the Oostanaula joined the Etowah, beneath the hills where Rome, Georgia would be built. From there the stream moved south on a circuitous course, gently bending in loops so large that later generations claimed it curved to touch every farm in the valley. For nearly one hundred and fifty miles, it flowed, deep and slow, navigable to Indian canoes, then frontiersmen's flats, and eventually steamboats. Only an occasional shoal exposed in low water, a fallen tree, or a drifting sandbar presented a problem for those who wished to use her.

Then suddenly the land fell away sharply, and the Coosa became another river. Wide Bends and slow reaches give way to nearly one hundred miles of rocks and reefs that old men in old times named the "Narrows," "Devil's Race," "Broken Arrow Shoals," "Hell's Gap," "Butting Ram," and finally, at the end, the "Devil's Staircase." The First Coosa dropped less than one hundred feet during its winding course; the second, flowing fewer miles, fell for times farther. From where it began to where it ended in the pool below the falls at the present site of Wetumpka, this Coosa was one of the wildest rivers in the Southeast.

----

*Head, Department of History and Foreign Languages, Jacksonville State University, Jacksonville, AL, 36265; phone 256-782-5632; hjackson@jsucc.jsu.edu.

Some twenty miles below the last fall the Coosa was met by its smaller sister stream, the Tallapoosa, and together they became the Alabama. The new river wound its way slowly west, along the northern boundary of what was later called the Black Belt, between bluffs that early explorers claimed made it look like a canal. Then, just past the high ground where Selma would be built, the Alabama met its last major tributary, the Cahaba, and turned south.

Now the landscape became more tropical. Cane and palmetto filled the swamps whose "dark-green recesses" created an atmosphere that one traveler compared to India and Ceylon. And finally the Alabama met the Tombigbee and both vanished into the river named Mobile. It was, I concluded, a fitting end, for it left the Alabama an Alabama river. It would share its name with the state and no other state could claim it.[2]

No sooner did men arrive than they began to alter the streams. Indians moved tons of rocks, constructed weirs to change currents and trap fish. They cleared landings, built river towns, and began the pollution process that continues up to today. Native Americans poisoned fish with buckeye pulp and devil's shoestring, threw village trash into the streams, and clouded the waters with runoff from their fields. But there were never enough Indians to do any long range ecological damage, so the waters flowed swift and clear when De Soto appeared, and they continued to flow that way when he departed.

We know little of what happened to the rivers during the two centuries after the first Europeans were gone. The few whites that came later left scant records, and archaeological evidence does not tell us much about river life. But in the late 17th century the English, French, and Spanish became interested in the region, and soon all three had outposts or allies in the Coosa-Tallapoosa valley.[3]

European competition for control of the river region eventually drew the Indians into conflicts in which they historically had no part. When the French and English fought, Native Americans served on both sides; and when Americans fought the English for their independence, each side had Indian allies. The English won the first war; the Americans won the second; the Indians lost both. Then the settlers came, citizens of the new United States, they floated the rivers and mucked their way along the Federal road. Settling, often illegally, wherever they could, these immigrants intermarried with Native Americans, which in time produced one of the most racially diverse populations in the new nation. As time passed, however, some of the Creeks, alarmed at the cultural damage they felt whites and mixed-breeds were doing to their nation, decided to drive them out. What followed is known in American history books as the War of 1812 and in Alabama history books as the Creek war. It could also be called "the War for the Alabama," for its outcome determined who controlled the river, farmed its bottoms, built towns and cities on its bluffs.[4]

And again, the Indians lost. Defeated in their river strongholds at Horseshoe Bend and the Holy Ground, the Creeks gave way and the Alabama became an American stream. This was confirmed by Major Howell Tatum, Andrew Jackson's engineer who surveyed the river in the late summer of 1814. Tatum quickly saw the potential of the river region, but at the same time he understood what flower would bloom there. "The rich lands on each side of the Alabama," he wrote in his journal, "are far superior to any that I have seen in any country, and I have no doubt [they] will prove a source of immense wealth to those who may hereafter be doomed to be the cultivators."[5]

So it was with the exhilaration that comes with victory and the fatalism that was part of the region's frontier Protestant heritage, the Age of the Alabama began. By 1819 the territory was a state, and a new capital -- Cahawba -- was built at the confluence of the Alabama and Cahaba. Meanwhile, upriver Selma was beginning on what was known as Moore's bluff, while near the forks of the Coosa and Tallapoosa two land companies decided to pool their resources and from this the town of Montgomery was born. Downstream, on the last high ground before the Alabama descended into the swamps and marshes, Claiborne rose.

To these and other sites more settlers came, and they brought with them a vision of what they wanted to create. What is remarkable is that in the brief time between settlement and Civil War, many came close to making that vision a reality. First they simplified eastern ideas and institutions, stripped them down to their essentials, so that government, church, and society could function as practically and efficiently as conditions required. Then, in time, as population and prosperity allowed, they began to elaborate on what they had created. They added offices and duties, paid heed to ceremonies and codes of conduct, and in the process left behind the simplicity of frontier farms for the civilized life of planter and merchant. And finally, the wealthiest and most successful among them set out to replicate the world of the wealthiest and most successful of their eastern counterparts by copying, as best they could, the life of the tidewater and low country planter. In short, they created the Old South but did it in so short a time that the frontier remained clear on the horizon behind them -- silhouetted but never out of sight.

The rhythms of the Alabama were the rhythms of the seasons, which complemented the rhythms of the cotton culture. In the spring, when supplies were needed from Mobile, the river was full and flowing; in the summer, when crops were laid by and thoughts turned to other things the river was low, slow, and ready to be fished or swam; but by late fall, when the cotton was ginned and baled for shopping the rains had raised it once again, and flats could slide easily over sandbars and gravel banks.[6]

What made it work so well, what brought it all together, was the steamboat. First it was the Harriett, which made the trip up to Montgomery in 1821, and before that decade ended the steam seemed full of those strange creatures.

To the modern mind tales of steamboats and steamboat travel are filled with the romance and adventure that folks associate with the Old South of myth and legend. On these 'floating palaces," so the story goes, Dixie aristocrats found circumstances and comforts like those they left behind on their great plantations, except that on the river the elements of their way of life were distilled into a finer, more elegant liquor. Along the steamboat's promenade deck southern ladies and gentlemen walked and talked with a style and grace found nowhere else in the hemisphere, while in the main saloon, riverboat gamblers waited for the planter rich enough, bold enough, and careless enough to make the trip worthwhile. Just as the modern mind continues to equate antebellum southern society with what is seen in the first few frames of Gone with the Wind, it continues to measure these vessels by images left over from the movie *Showboat* and from modern replicas that ply southern rivers.

Much of this, of course, is myth, but we know that a myth persists because, whatever its flaws, in ways it is true. For most passengers a journey on an antebellum steamboat *was* an adventure so exotic, so filled with wonder and excitement that the line between reality and romance blurred, and the distinction lost any real meaning. Everyone who was part of the scene, from deckhand to passenger to the captain himself, was transformed into something else, something greater than he would have been on land.

Little wonder, therefore, that there are those among us who thoughtlessly cling to the dream of those glory days with the same rose-tinted tenacity as they hold to the illusion of an Old South of lordly gentlemen and lovely ladies set in manorial splendor and served by numerous, happy slaves. And because it is so attractive, it is little wonder that so many others, knowing better, also prefer the myth. I thought long and hard about this as I began to write about the boats; and concluded, why labor the point? As one of Selma's loveliest ladies told me on a cloudless October day as we sat by the river at Old Cahawba, "if it did not happen that way, it should have."[7]

Still, I was forced by fact to point out that not all steamers were floating palaces. More often than not they were utilitarian craft like the *Lula D* -- "Not one of those high-living boats," according to the nephew of its owner; its appearance reflected that of its captain, a man "of the pioneer type" who was not inclined to waste time or money on that "grand style and princely fare" that attracted discerning passengers. "Not fastidious about his eating," the captain was rumored to have "assigned his poorest deck hand to the cooking task." The job was no easier (and the food no better) because in those "pre-ice days the beef bought at landing stops soon became too strong and the eggs were that way" from the start. So unappealing was the food on the boat that crewmen once scraped the sides of empty molasses barrels for sugar to give flavor to the "saucer sized biscuits" the chef produced.

On the other end of the scale from the *Lula D* was the "elegant little *Mary*" commanded by its "Catholic Captain" the young John Quill, a ship noted for "the gentlemanly tone and manners of all the officers" whose conduct was such that it was reported that "during all the busy and exciting labor of cotton loading there was not heard a single oath, not so much as one rude word to shock a lady's ear." So confident was Quill of the conduct of his crew that he invited ministers to travel free.

The *Mary* was unique, as was Captain Quill. Most boats and crews were probably like the one on which a travelers reported being kept up all night by a "profane mate" who at least made a concession to the fairer sex and only cussed after dark, when the fairer sex was asleep. On most boats gambling was accepted, bars were common features, and one passenger reported that "an intoxicated man on board was no uncommon sight." These boats were no place for a man of the cloth, unless he was seeking unconverted souls.[8]

Rivermen, like John Quill, quickly became central to my story, but it was in the infrequently told tales of the laborers, deckhands, roustabouts, and stevedores, that I found an attachment to the streams that went beyond the ownership of a boat.

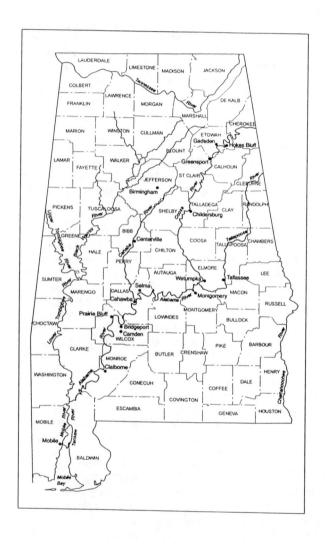

Towns of the Alabama River System - *Rivers of History: Life on the Coosa,
Tallapoosa, Cahaba, and Alabama*
(Courtesy of University of Alabama Press)

Issam Morgan, a former slave, was one of these. Still a boy when the war ended, he stayed on the plantation until he was twenty-one and might have stayed longer had his "Massa" not come one day and simply told him, "Issam, you is a grown man now. You is got to boss your own business . . . . I can't keep you no longer." Massa wished him good luck, assured him that he would make somebody a good worker, and sent him on his way.

With that Morgan joined the ranks of landless freemen who had to make do in a world over which they had little control. Issam Morgan headed for the river. In the years that followed, he recalled how he "worked different jobs, sich as loader, roustabout on different steamboats an' [in harvest season] cotton picker." Interviewed by the WPA in the 1930s, he remembered the vessels, and "one of de ole songs sang on de boats":

> De *John T. Moore*
> De *Lula D*
> An' all dem boats is mine.
> If you can't ship on de *Lula D*
> You ain't no man o' mine.[9]

On shore there were the towns. In studying them I discovered that they were not always what they seemed. Cities like Selma, Montgomery, and (later) Gadsden each had a special bond with the stream on whose banks they sat, but none of these were purely river towns. Each became more, and if they had not, they would never have survived and prospered.

Consider Prairie Bluff. Born in the 1830s of a booster's dream of profit, Prairie Bluff came into being when two real estate syndicates bought the land, divided it between them, and then subdivided it for sale. Located in the heart of the Black Belt, the town soon had warehouses to store cotton and a long slide to carry it down to the river. In time there were some twenty stores, two wagon factories, a gin factory, several blacksmith shops, shoemaker shops, and a large confectionery. It also boasted many barrooms, billiard tables and ten pin alleys, as well as a local racetrack. There were also a half-dozen practicing physicians to patch up the wounded after barroom brawls and a number of lawyers ready to file suit for the survivors. What was lacking, you might have noticed, was a church.

Like so many river settlements, Prairie Bluff's foundation was weak. Each citizen was a speculator at heart; so he expected to sell his improvements and move on. Because of this attitude, residences and business establishments always had a temporary look that gave the whole settlement a rude and makeshift appearance. So it followed that when the economy turned down, as it seemed to do every 15 to 20 years, conditions in the town appeared worse than they actually were. New arrivals took one look and moved on. Soon, residents left as well. In the late 1830s a financial panic followed by a yellow fever epidemic set Prairie Bluff's decline into motion. It never stopped. Other towns rose to take its place, and though it held on through the Civil War, before the end of the century the site was deserted.[10]

By the 1850s the Alabama River valley was feeling flush. The seat of government had been moved to Montgomery and the Legislature was meeting in the new capitol that a northern visitor described as a "true Athenian Yankeeized structure [in] this novo-classic land, erected on a site worthy of a better fate and edifice." As for Cahawba, though legend has the town declining after the government was moved, in truth it recovered quickly and by mid-century was a thriving social and economic center. Even Selma was on the rise, though like other landing-towns, it still retained its frontier image. Governed by a class of men who would, on the same night, pass an ordinance prohibiting cock-fighting and then, according to council minutes, adjourn to take a drink, the "queen city of the Black Belt" (as boosters would later call her) sent a mixed message to those who might settle there.[11]

Then the war came.

If secession caught anyone in the river region by surprise, they probably weren't paying attention. And when it came, it came swiftly. Walter Calloway, a young slave on a Montgomery County plantation, remembered how his brother came back from the capital and told them "hell done broke loose in Gawgy." When they questioned him closely they found out "dat some of the big mens called a meetin' at the capitol on Goat Hill . . . an' done busted the Nunited States wide open."[12]

In the months that followed, all along the river young men rallied to the Confederacy. Selma raised five companies, whose membership reflected the diversity of the growing town. There were the socially prominent "Magnolia Cadets," in whose ranks were "the first young men of the place," the "Selma Blues" made up of the "more sober, settled men of the city," and the "'Phoenix Reds' composed almost entirely of working men." Before it was over Selma would send 600 soldiers into the Army.[13]

Downstream at Cahawba, in a ceremony reenacted in towns and villages throughout the state, scions of many a Black Belt family joined the local company and prepared to leave. Just before they departed their commander, Capt. Christopher C. Pegues, accepted a banner from a local belle, and swore to bear it to "victory or to death." "Right royally," remembered Anna Gayle Fry, "was that oath fulfilled." Pegues proved an able officer, became a colonel, and might have been a general and later a governor or a senator, had he not fallen leading his men at Gaines' Mill, in the summer of '62.[14]

But that April day at the confluence of the Alabama and Cahaba, war was far away for Anna Fry's "wealthy, cultured, young gentlemen who [like Pegues] turned their backs upon luxuries and endearments of affluent homes." It was still ahead for Selma's workers who left jobs and families to join the cause; it was still ahead for "the more sober settled men" who should have known better but enlisted anyway.

It also lay ahead for scores of farm boys from a region of farm boys, whose families counted their acres in tens, not thousands, who owned no slaves, and who had never seen a Yankee until they shot at one. They enlisted (and were later conscripted) without ceremony. No flags were consecrated for them, no bands played, no belles gave them flowers and kisses.

It also lay ahead for men like "massa Cal" from Lower Peachtree, below Prairie bluff, who understood the slogging horror of war and went anyway. But his ceremony of departure was different, and more personal. Before leaving, "massa Cal" summoned his slave Cato, who through one of those unions that white southerners could neither acknowledge nor ignore, was also his kinsman. He told him (according to Cato): "you's always been a 'sponsible man, and I leave you to look after the women and the place. If I don't come back, I want you always to stay by Missy Angela!"

"Fore god, I will, massa Cal," Cato replied.

"Then I can go away peaceable," the planter said, and he left.[15]

Had "Massa Cal" and the others stayed at their homes on the Alabama, the Coosa, the Tallapoosa, and the Cahaba, it would have been years before the war came to them, if it came at all. So they went to the war and many, so very many, never returned.

Of all the river towns and cities it was Selma that gained the most from the conflict. Early in the struggle, chief of ordnance, General Josiah Gorgas, began turning the town into a major industrial center, and in time the complex included a naval yard and naval ordnance works, along with foundries, mills and factories that employed as many as 10,000 workers -- more than the entire population of the whole county before the war. Selma also lost the most from the conflict, for in the spring of 1865, Union cavalry under General James Wilson invaded Alabama, and when they finished their work Selma was in ruins. Montgomery avoided a similar fate by surrendering without firing a shot.[16]

But in the years between, as Selma industries boomed and factory managers attempted to meet military needs from an ever-dwindling supply of raw material, a literary contest ensued that took Selmaians' minds off the struggle -- at least temporarily. It began in the fall of 1863 when John Harrolson, local agent for the Nitre and Mining Bureau, discovered that a shortage of nitre threatened gunpowder production. Realizing that urine was a natural source of the necessary object, Harrolson put the following announcement in the *Selma Sentinel*:

> The Ladies of Selma are respectfully requested to preserve
> all their chamber lye collected about the premises, for the
> purposes of making "Nitre." Wagons with barrels, will be
> sent around for it by the subscriber.

Thomas B. Wetmore, a friend of Harrolson, was inspired by the notice, and soon this poem was being circulated through the town.

> John Harrolson! John Harrolson!
> You are a funny creature;
> You've given this cruel war
> A new and curious feature.
> You'd have us think, while every man
> Is bound to be a fighter,
> The Women (bless the pretty dears),
> Should be put to making nitre.
>
> John Harrolson! John Harrolson!
> How could you get the notion

To send your barrels 'round the town
To gather up the lotion.
We think the girls do work enough,
In making love and kissing;
But now you'll put the pretty dears
To patriotic pissing.

John Harrolson! John Harrolson!
Could you not invent a meter,
Or some less immodest mode
Of making our salt-petre?
The thing, it is so queer, you know--
Gunpowder like and cranky--
That when a lady lifts her shift
She shoots a bloody Yankee!

John Harrolson! John Harrolson!
Whate're was your intention,
You've made another contraband
Of things we hate to mention,
What good will all our fighting do
If Yanks search Venus' mountains,
And confiscate and carry off
These Southern nitre fountains.[17]

The poem was eventually sent to soldiers who passed it around until it was found in all theaters of the war and on both sides. Today it ranks among the best-known "underground" literature of the conflict.

With the war's end river enthusiasts began laying plans to install a locking system to conquer the rapids of the Lower Coosa and open the river from Wetumpka to Rome. Meanwhile, down on the Alabama, commerce continued--indeed increased--for all the "Old South" mythology notwithstanding, the "Golden Age of Steamboats" was after the war, not before. Four locks were built on the Upper Coosa, where the rapids begin, and another was built at the end of the run, just below Wetumpka. But soon after the turn of the century, engineer assessments concluded the project would never pay for itself, so efforts to open the Coosa were abandoned.[18]

But not for long. Into the breech stepped William Patrick Lay and his newly formed Alabama Power Company, and soon he and his partners were "putting the loafing streams to work." Between 1911 and 1929 the Alabama Power Company built four major hydroelectric dams – Lay, Mitchell and Jordan on the Coosa, and Martin on the Tallapoosa. In the process they changed the lives of valley folks. Soon cities and towns were lit by electricity; there were radios and refrigerators in homes that could afford them; industry was more widely disbursed; and farmers far from the rivers had employment alternatives never available before.[19]

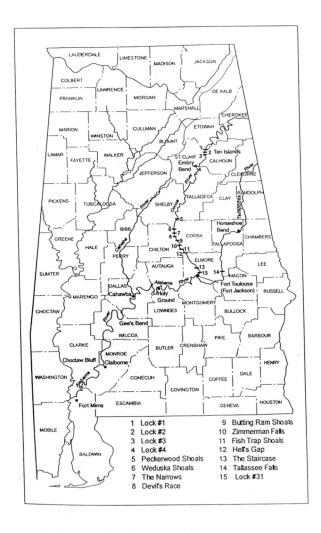

Landmarks on the Alabama River System - *Rivers of History: Life on the Coosa, Tallapoosa, Cahaba, and Alabama*
(Courtesy of University of Alabama Press)

The dams also changed the rivers. The great falls at Tallassee were gone. Lake Jordan covered Welonee Creek Reef, Hell's Gap, Fish Trap Shoals, and most of the Devil's Staircase. Mitchell Dam sat atop Duncan's Riffle; and the water behind it rose over the Devil's Race. Meanwhile Lay Lake washed over the Narrows, Weduska and Peckerwood Shoals. Gone were the islands and ferries, named after their people – Johnson, Smith, Knight, Pate, Sauley, House, Gray, Noble, Thompson, Higgins, Adams, and a score of others – a roll call of the region's pioneers lost beneath the waves. The "loafing streams" had been put to work, and this was the price Alabama paid.

As I cataloged these changes I began to think back on the ecological alterations the river had undergone. As settlement increased along the waterways there were fewer references to the clear green streams and more to rivers that were, according to one traveler, "the color of chocolate and milk." By the 1850s historian Albert J. Pickett was complaining how the "waters of Alabama begin to be discolored; the forests have been cut down; vast fields of cotton, noisy steamers, huge rafts of lumber, towns reared for business . . . mills, factories and everything else that is calculated to destroy the beauty of a country and to rob man of his quiet and native independence, present themselves to our view."[20]

It got worse after the war.

Maybe just because they were keeping better records or because there were more people around to remember, but now floods seemed to come more frequently and fiercely. In '65, '66, '72, '86, '88, and '92 the rains fell, and the rivers rose, fast, for the land could not contain the water as it had before. Hardscrabble farmers and debt-ridden planters had clear-cut any land that would grow cotton, and fields ran down to the creeks and out to the edge of river bluffs. So it followed that in the spring, before the crop was up, the bare earth soaked in what rain it could and rejected the rest. The runoff, heavy with the soil it was supposed to nurture, flowed into the river, and the river, after taking what water it could, turned the rest back on the land.

Rivermen saw the results. When the silt that colored the water settled out, it clogged the channel and hindered navigation. What they did not see was the beginning of an ecological alteration that continues to threaten the rivers today. As the land wore down, farmers turned to fertilizers to replenish the soil. These supplements washed away as well, and in the streams they began to work changes on vegetation that would not be realized for decades.

But at the time, what was there to notice? Schools of fish still came upriver to the Wetumpka shoals to spawn, sturgeon were frequently caught in the Cahaba, and catfish thrived no matter what happened to the streams. There was no time for concern – not yet.

Biologist Herbert H. Smith knew the richness of Alabama's rivers. During the first decade of the 20th century he collected thousands of snails and mussels from the Coosa and left a record of a river where a scientist could gather 150 specimens in a space not twenty yards long. He found reefs and shoals, small channels, and backwater pools crowded with life. Before he finished Smith confirmed that the Coosa River had "more endemic molluscan species than any other stream in North America."[21]

Less than ½ a century later they were gone. Hydroelectric dams had turned miles of reefs and formerly quiet reaches between rapids into silt-accumulating, habitat destroying lakes. A study of the same Coosa in the 1940s concluded that the river had been so "offishly treated" that of the twenty-four fresh-water snails that the American Fisheries Society listed as believed to be extinct, **all** were from Alabama. "No other state," according to the Society, "is suspected of killing off an entire snail species." At least seventy more Alabama snails and mussels were listed as threatened or endangered.[22]

The dams inadvertently caused another problem. In the past cities and factories dumped their refuge into the streams for it was generally assumed that "the solution to pollution is dilution." But the dams stopped the rivers and turned the streams into large holding ponds. When the navigation dams on the Alabama and the rest of the Power Company dams on the Coosa were completed in the 1960s and 1970s, the rivers ceased to be rivers at all. Now they were a series of elongated lakes. The dams had created a whole new ecosystem.

But even before the dams, old river ways were slipping away. Packet steamers like the *Mary* assured passengers they would arrive in the morning, afternoon, or evening, but New South businessmen demanded a more precise schedule. The railroad siphoned these passengers away, and as the boats lost that important source of revenue their owners let the vessels become shabbier and shabbier, which drove away what few riders remained. By the early 20th century those steamboats still on the river were little more than cotton barges. Then the boll weevil took the cotton and there was nothing left. In 1932 the *Helen Burke*, the last of the great Alabama River packets, was tied up to the bank of the Mobile River and stripped of her machinery. Soon her seams split and she slowly sank to the bottom. She was the last of the breed.[23]

Up on the Coosa it was much the same. Roads and railroads became the major means of transportation and the river fell into disuse. One by one the locks were abandoned. Finally, in 1939, a new bridge was opened south of Gadsden to serve travelers on the highway that ran from North Alabama to Montgomery. It was a turnbridge, and at the dedication it was cranked open to show how the span could be moved to let boats through. Then it was cranked back into place and never opened again. There was no need. There were no boats.[24]

Citizens of the river region survived the Depression, prospered during World War II, and after the conflict set about to "develop" the rivers. Instead of steamboats, landings, and wharves, they built more hydroelectric dams on the Coosa and Tallapoosa, pushed the Corps of Engineers to construct navigation locks on the Alabama, and sniffed the air and smelled progress as paper mills sprung up along the banks. They put into motion the forces that would, as one of them promised, "fence the river with smoke stacks."[25]

But it didn't work that way. Today, hydroelectric lakes have become recreational paradises instead of industrial basins, lined by vacation homes and marinas instead of factories. Moreover in an ironic way the paper industry has actually preserved much of the rural way of life that has always been characteristic of the river region – especially on the Alabama.

Down there the rhythm of life was never the rhythm of the factory, and though the paper mill does force locals to accept shift work and routine, it does little else to alter their existence. While workers arrive and leave according to schedule, when shifts change they do not make their way to a nearby factory town, the sort often associated with mills. Instead they get into their pick-up trucks and as fast as the law allows (sometime faster), they head home to towns and farms where many of their families have lived for generations. For some the trip takes an hour or more, and woe be to the unsuspecting motorist who happens to be driving by when they roar out of the gate.

In a sense this commute reflects a regional determination to hold to a way of living that is usually threatened, if not destroyed, by industrialization. But one must understand that the world they speed home to survives because of the mills, not in spite of them. If the forest product industry is to succeed it must have forests, so it follows that paper mills promote and praise rural life and values. Apart from already established towns like Selma, Montgomery, and Gadsden, the urbanization usually associated with industrialization has not occurred along the rivers, and folks there like it that way.

So it follows that the paper mills, though they get much of the blame for polluting the rivers, are hardly the only, or the worst, culprits. In the fall of 1990 two men canoed the Coosa-Alabama from near Rome to Mobile Bay. On the trip they passed mines of various sorts, some seventy municipal sewage treatment plants, hundreds of farms and timber operations, and scores of industries, large and small, only five of which made paper. All these polluted the river in some way, and if the river is to be cleaned up, all will be affected.

One example brings the problem into focus. By the end of the 1980s sections of the Alabama and the Coosa were so polluted that their capacity to accept more discharge was all but exhausted. In other words, unless the rivers were cleaned up, existing industries and municipalities could not expand and new industries could not come in. Faced with this businessmen began to call for an increased effort to restore the streams. Conservationists welcomed these allies, even though they knew that their concern did not reflect a new environmental awareness. Industrial promoters wanted to reduce pollution to make way for more pollution. Such is the irony of environmental politics in Alabama.[26]

What the canoeists did not see were other boats. Except for a few fishermen and an occasional barge, they had the river to themselves. In 1960, a report on America's inland waterways found "no traffic of consequence on the Coosa-Alabama system." Alabamians, working through the Coosa-Alabama River Improvement Association, argued that this would change if navigation dams were built on the Alabama, and after a furious lobbying effort, the money was allocated and three lock and dam combinations were constructed. Then, in the 1970s, the oil crisis gave ammunition to those who wanted to make the whole system navigable from Mobile to Rome. Pointing out that each Alabama Power Company dam included in it an earthen section for a lock, advocates pressed Congress to provide the money to open up the Coosa. For a moment it looked like it might happen, then a cost-benefit report revealed that it would take nearly 100 years for the project to pay for itself. With that, the Corps of Engineers was directed "to undertake an orderly suspension" of the plan."[27]

The Coosa-Alabama River Improvement Association vowed to keep pressure on Congress, and it has, but to no avail. Today, except for occasional barges loaded with sand and gravel, some fuel oil, and maybe a piece of machinery that is too big for the highway, not much is moving along the Alabama. And nothing is moving on the Coosa. Although these streams continue to serve the state and the region in many ways, they are no longer commercial carriers. And, unless some national emergency redirects national priorities, it is unlikely they will ever be again.

---

[1] "Coosa-Alabama River Improvement Association, Biannual Newsletter," Spring/Summer, 2002. This article is based on Harvey H. Jackson, III, *Rivers of History: Life on the Coosa, Tallapoosa, Cahaba, and Alabama* (1995).

[2] Harriet Martineau, *Society in America*, 2 vols. (1837), 2:226. Early accounts of the rivers can be found in William Bartram, *Travels*, ed. Francis Harper (1791, reprint, 1958); C. L. Grant, ed., *Letters, Journals, and Writings of Benjamin Hawkins, 2 vols.* (1980). A good early map of the upper part of the river system, showing all the rapids and ferries is COOSA RIVER, GA. AND Ala.; reported in Accordance with H. Doc. 308/69/1, prepared by the Corps of Engineers, December 31, 1928, in the Alabama Department of Archives and History, Montgomery, Alabama.

[3] The debate over where De Soto went and what he did will probably never be settled. However, my impressions of the expedition and its impact have been shaped by reading the work of Charles Hudson, most recently *Knights of Spain, Warriors of the Sun: Hernando de Soto and the South's Ancient Chiefdoms* (1997). Hudson, *The Southeastern Indians* (1976) is an excellent source on Native Americans in the region.

[4] Recently two major works have appeared that add much to the study of the Creeks. See Kathryn E. Holland Braund, *Deerskins and Duffels: Creek Indian Trade with Anglo America, 1685-1815* (1993) and Claudio Saunt, *A New Order of Things: Property, Power and the Transformation of the Creek Indians, 1733-1816* (1999). For an excellent account of land travel into the region see Henry deLeon Southerland, Jr., and Jerry Elijah Brown, *The Federal Road through the Creek Nation and Alabama, 1806-1836* (1989).

[5] John Spencer Bassett, ed., "Major Howell Tatum's Journal, while acting Topographical Engineer (1814) to General Jackson, Commanding the Seventh Military District," *Smith College Studies in History* 7(October 1921 to April 1922).

[6] The standard study of early settlement remains Thomas P. Abernethy, *The Formative Period in Alabama, 1815-1828* (1922). For a list of other books, articles, and primary sources relating to the period see Jackson, *Rivers of History*, 287-290.

[7] Reading travel accounts of the period, especially those written by James Silk Buckingham, George W. Featherstonhaugh, Basil Hall, Margaret Hunter hall, Charles Lyell, Harriet Martineau, Tyrone Power, and James Stuart, under score the excitement (and often the discomfort) of river travel. A full listing of these works may be found in Jackson, *Rivers of History*, 289-290. The insightful Selma belle was noted Alabama author and my cousin-in-law, Kathryn Tucker Windham.

[8] Stories of John Quill, the *Mary*, and the *Lula D* come from the *Morning Star and Catholic Messenger*, May 29, 1901; and the *Montgomery Advertiser*, August 17, 1875 and September 21, 1949.

[9] Accounts of African Americans working on the river are found in George P. Rawick, ed., *The American Slave: A Composite Autobiography* (1941), which contains slave narratives compiled under the auspices of the WPA during the New Deal era. Vol. 6 relates directly to Alabama. See also E. A. Botkin, ed., *Lay My Burden Down* (1945).

[10] For an account of river towns see Jackson, *Rivers of History*, 59-72.

[11] William Howard Russell, *My Diary, North and South* (1861), 167-168; John Hardy, *Selma: Her Institutions and Her Men* (1879), 14, 66. See also, Alston Fitts, III, *Selma: Queen City of the Blackbelt* (1989).

[12] Rawick, *The American Slave*, 6:52-53.

[13] Hardy, *Selma*, 46.

[14] Anna M. Gayle Fry, *Memories of Old Cahaba* (1908), 30-35, 105.

[15] Botkin, *Lay My Burden Down*, 87.

[16] For more on Alabama cities during the Civil War see James Pickett Jones. *Yankee Blitzkrieg: Wilson's Raid through Alabama and Georgia* (1976); Arthur w. Bergeron, Jr., *Confederate Mobile* (1991); and William Warren Rogers, *Confederate Home Front: Montgomery during the Civil War* (1999).

[17] Heavily censored versions of these verses have appeared in a number of publications. The full poem can be found in William Moss, ed., *Confederate Broadsides: An Annotated Descriptive Bibliography* (1988), 155, and Walter M. Jackson, *The Story of Selma* (1954), 200-201. Jackson's book also contains some of the responses to the poem.

[18] For an account of this post-Civil War activity see Jackson, *Rivers of History*, 140-156.

[19] For information on the early years of the Alabama Power Company see Harvey H. Jackson III, *"Putting Loafing Streams to Work": The Building of Lay, Mitchell, Martin and Jordan Dams, 1910 to 1929* (1997).

[20] Albert James Pickett, *History of Alabama and Incidentally of Georgia and Mississippi, from the Earliest Period* (1851), 311.

[21] Calvin Goodrich, "Certain Operculates of the Coosa River," *Nautilus* (July, 1944), 1-3.

[22] Ibid., 1.

[23] Capt. Simon Peter Gray to D. L. McCall, October 9, 1956, in Steamboat File, Alabama Department of Archives and History, Montgomery, Alabama.

[24] Mrs. Betty Sue McElroy of Gadsden told me the story of the turnbridge.

[25] For a more detailed account of attempts to industrialize the river region, see Jackson, *Rivers of History*, 223-240.

[26] In December, 1990, the *Alabama Journal* published a series on pollution of the rivers that was subsequently reprinted in a special edition entitled "Alabama's Rivers: An Endangered Resource." This section is based on those articles.

[27] The almost opening of the Coosa is discussed in Jackson, *Rivers of History*, 220-222.

Initial History and Strategic Planning of the Environmental & Water Resources Institute (EWRI)

Conrad G. Keyes, Jr.[1], ScD, P.E./P.S., F. ASCE, F. NSPE
Jerry R. Rogers[2], Ph.D., P.E., F. ASCE

Overview

Over a period of years from the prior Management Group and Technical Activities Committee structure, ASCE has created new National Institutes: Structural Engineering Institute (SEI), Geo-Institute (G-I), Architectural Engineering Institute (AEI), and in 1999 the Environmental and Water Resources Institute (EWRI). Additional institutes added in recent years are Coastal, Ocean, Ports, and Riverine Institute (COPRI), Construction Institute (CI), and the Transportation & Development Institute (T&DI) in 2002.

The ASCE Task Committee to Create the EWRI was formed in fiscal year 1997 and for two years chaired by Michael L. Stevens/Conrad G. Keyes, Jr. to plan the transition for the Environmental and Water Resources Institute from three ASCE Divisions: Environmental Engineering, Water Resources Engineering, and Water Resources Planning & Management plus the Water & Environmental Standards Council. In July 1999, the ASCE Board of Direction approved the formation of the EWRI, to officially begin October 1, 1999.

Founding Brief History

Created in 1999, the Environmental & Water Resources Institute (EWRI) is a semi-autonomous institute of the American Society of Civil Engineers (ASCE), the country's oldest national engineering society. EWRI's services are designed to complement ASCE's traditional civil engineering base, and to attract new categories of members (non-civil engineer allied professionals) who seek to enhance their professional and technical development.

[1] Past President, EWRI, P.O. Box 1499, Mesilla Park, NM 88047, and Professor & Dept. Head Emeritus, Civil & Geological Engineering Department, New Mexico State University, Las Cruces, NM 88003 (505-523-7233) (mailto:cgkeyesjr@zianet.com)
[2] Civil & Environmental Engineering Department, University of Houston, Houston, TX 77204 (713-743-4276) (mailto:djrogers@pdq.net)

EWRI was formed in keeping with the Strategic Plan of ASCE to provide products and services to a variety of engineering teams working in technical, educational, and professional areas. Former ASCE Technical Division enrollees (Environmental Engineering, Water Resources Engineering, Water Resources Planning and Management - as well as the Water and Environmental Standards Council), form an EWRI membership base of 20,000 individuals.

Vision. The EWRI will be a recognized worldwide leader within ASCE for the integration of the technical expertise and public policy into the planning, design, construction and operation of environmentally sound and sustainable infrastructure impacting air, land, and water resources.

To accomplish this vision, EWRI commits to:
• A diverse and empowered membership
• Excellence in products and services
• Collaborative partnerships
• Innovative programs and solutions

Mission. EWRI will provide for the technical, educational and professional needs of its members, and serve the public in the sustainable use, conservation, and protection of natural resources and the enhancement of human well being by:

• Advancing the knowledge and improving the understanding of relevant sciences
• Improving the practice of engineering
• Partnering with governmental, industrial, educational and international organizations.

Summary of Initial EWRI Governing Board, Newsletter, Journals, EWRI 2001 Congress, ASCE Annual Conferences, & Website

The EWRI Task Committee planned EWRI sessions at the Charlotte Annual ASCE Conference & Exposition (coordinated by Norman L. Buehring) with a Founding Ceremony & Reception October 19, 1999 (coordinated by James R. Groves). The Founding EWRI Governing Board included President Conrad G. Keyes, Jr., President Elect Jeffrey B. Bradley, Vice President Cecil Lue-Hing, Michael A. Ports (representing Products & Marketing area), Jerry R. Rogers (appointed by ASCE BOD), Robert C. Williams (representing Member Services area), Darell D. Zimbelman (representing Technical Activities area), and EWRI Treasurer Thomas M. Rachford (non-voting member). The EWRI/ASCE Staff Contacts John E. Durrant/Brian Parsons (both now Institute Executive Directors), Elizabeth Downey and Autumn Richter provided the Staff assistance. Jean A. Bowman of College Station, TX was the first EWRI "CURRENTS" Newsletter Editor. All journals of the Divisions were continued and an

EWRI magazine was planned. Besides administrative groups, there were the three major EWRI areas: Member Services, Products & Marketing, and Technical Activities, which included the Standards Council. The first EWRI Congress (World Water & Environmental Resource Congress) was held on May 20-24, 2001 at the Orlando Clarion Plaza, Orlando, FL (steering committee chaired by Kyle E. Schilling) with a Specialty Symposium on Integrated Surface & Ground Water Management (chaired by Bijay Panigrahi). The ASCE Annual Conferences with EWRI Coordinators included: Donald Phelps- Seattle- October 19-21, 2000; Michael A. Ports- Houston- October 10-13, 2001; and A.J. Fredrich- Washington, D.C. -2002. The EWRI Bylaws and Committees are listed on pages 380-418 of the *ASCE Official Register 2002*. The EWRI website is: http://www.EWRInstitute.org.

Founding History of Leadership

A seven-member Governing Board administers the affairs of the EWRI of ASCE. The 1999-2001 Founding President was Conrad G. Keyes, Jr., Sc.D., P.E./P.S., F.ASCE, and F. NSPE. The 2001-2002 President is Jeffrey B. Bradley, Ph.D., P.E., and F.ASCE. The first EWRI Executive Director (ED) was John E. Durrant, P.E. and the present EWRI ED is Brian Parsons, P.E. In addition to Governing Board Level Committees, EWRI is organized around three primary interest areas – Technical Activities, Products, and Member Services.

Technical Activities. The EWRI technical activities featured seven councils and more than 100 active task committees working on publications such as manuals of practice and technical standards, as well as providing commentary on public policy. Featured councils address engineering issues such as planning and management, infrastructure, irrigation and drainage, watersheds, hydraulics and waterways, cross media (environmental issues), and standards development.

Products. EWRI products, typically developed by the technical activity areas, are coordinated and marketed through the efforts of the Products Councils. EWRI produces five technical journals, develops and hosts several annual technical conferences, produces a host of engineering standards, and engineering practice publications, and encourages continuing education courses.

Member Services. EWRI membership area councils encourage and promote individual and organizational memberships. The councils plan and conduct activities to retain current members, encourage member participation in Institute activities, and administer to student activities, foster international cooperation, and facilitate interaction with ASCE sections and branches.

Proposed Liaisons with ASCE Sections/Branches

The initial EWRI Section/Branch Activities Council (Chair Karen C. Kabbes) was asked to foster and promote local activities in the environmental/water areas via the ASCE network. For example, the ASCE Texas Section and its Environmental, Hydraulics and Water Resources Groups and its 17 Branches were asked to actively seek formal links with EWRI and exchange communications with the EWRI Section/Branch Council. See the EWRI Strategic Plan 2000 below (1.4- Establish liaison mechanism with Sections and Branches).

Initial Membership, Dues, and Strategic Plan

ASCE members can join one Institute free such as EWRI by simply calling ASCE 1-800-548-ASCE or contacting the ASCE website: http://www.asce.org. The annual dues are $20 if EWRI was an additional Institute. A non-ASCE member may join EWRI for $85 Membership fee. There are also EWRI Organizational Member opportunities.

The 2002 Strategic Plan of the EWRI included Goals, Objectives, and Action Items that were developed by a group of 20-25 of the EWRI Leaders in January 2000 in San Antonio. This second strategic planning session of the EWRI groups was conducted at the same time as the beginning of the recent ASCE Strategic Plan under consideration in 2000. The details of that EWRI Strategic Plan 2000 is provided below.

PROPOSED EWRI STRATEGIC PLAN 2000
GOALS, OBJECTIVES, AND ACTION ITEMS

Goal 1: Expand and diversify the membership base. Member Services EXCOM			
	OBJECTIVE	ACTION	COMMITTEE/ COUNCIL
1.1	Increase membership by 2000 by 2003 (2)	1.1.1 Execute the Membership Development Plan.	Membership Development & Retention Council
1.2	Increase younger member involvement*	1.2.1 Establish formal younger member "forum", "activities," etc. (4)	Membership Development & Retention Council
1.3	Increase student involvement*	1.3.1 Double student participation in EWRI activities (3)	Student Activities Council
1.4	Establish liaison mechanism with sections and branches. (5)	1.4.1 A general Institute protocol for interaction with Sections and Branches will be developed.	Section & Branch Activities Council/Staff
1.5	Conduct member survey for products and services (6)	1.5.1 ASCE is planning on doing a market research campaign and it is suggested that EWRI use this vehicle at a nominal cost to perform this research for EWRI. John Durrant will investigate the option of a membership consultant.	GB/Special TC
1.6	Develop four continuing education courses (7)	1.6.1 Consult with Technical Activities EXCOM and P&M EXCOM to determine state-of-the-art topic areas for continuing education courses as a benefit to EWRI members.	In collaboration with Continuing Education Council
1.7	Place all "products" on web (9)	1.7.1 Develop online reference for EWRI products.	In collaboration with Products & Marketing EXCOM//Staff
1.8	Enable web communications (10)	1.8.1 Develop online EWRI membership directory.	Communication Council/Staff
Goal 2: Partnering with external Groups. Governing Board			
	OBJECTIVE	ACTION	COMMITTEE/ COUNCIL
2.1	Identify worldwide collaborative partnerships, both internally and externally*	2.1.1 Establish worldwide collaborative partnerships (8), and develop partnering agreements with specific product goals.	Governing Board, EOCC

	**Goal 3: Improve the advancement and transfer of technology.** **Technical Activities EXCOM**
	OBJECTIVE		ACTION	COMMITTEE/ COUNCIL
3.1	Develop more timely and integrated publications appealing to a broader audience and resulting in increased sales. *	3.1.1	Each TA Council to participate in obtaining paper submittal to EWRI Journals	Technical Activities EXCOM in collaboration with Publications Council
		3.1.2	Identify new target audiences for new and existing publications (Journals, MOPs, Standards)	
3.2	Develop timely and targeted conferences and workshops that result in increased attendance and Conference exhibitors. *2000 – Kansas City, Minneapolis, Ft. Collins, Seattle; 2001– Orlando, Houston; 2002 – 3 needed + DC Convention; 2003 –EWRI Congress	3.2.1	Each council with TAC will sponsor at least 1 track at the bi-annual EWRI conference.	All Technical Activities committees/ Councils
		3.2.2	Each Technical Committee is suggested to be responsible for at least one session per track.	
3.3	Develop more standards and manuals of practice* - 5 Standards needed per year; 1 MOP needed per year. *	3.3.1	Streamline the standards development process and accelerate the production of standards for more technologies, materials, and processes.	Publications Council in collaboration with Standards Development Council
3.4	Identify members' needs and develop new programs and services to meet these needs. *	3.4.1	Conduct member survey for products and services. (6)	Technical Activities EXCOM in collaboration with Member Services Area
3.5	Develop continuing education programs that promote lifelong learning. *	3.5.1	Develop four continuing education courses. (7)	Continuing Education Council
3.6	Encourage and reward innovative programs. *	3.6.1	Establish EWRI awards mechanisms (11) to provide appropriate recognition and participation for all disciplines within EWRI and the related professions.	Awards Committee in collaboration with Technical Activities Awards Subcommittee
3.7	Enable web communications. (10)	3.7.1	Place "products" on the web, expand EWRI website delivery system and improve information transfer for continuing education, publications, and specialty conferences through the use of state-of-the-art communication technology and other resources. (9)	Emerging Technology Council in collaboration with Technical Activities EXCOM

Goal 4: Strengthen EWRI's public policy role. Products & Marketing				
OBJECTIVE		ACTION		COMMITTEE/ COUNCIL
4.1	Implement procedure for formal public involvement and policy development. *	4.1.1	Increase the quantity and quality of EWRI products.	Products & Marketing EXCOM
4.2	Mitigate and respond to regional and national disasters and emergencies. *	4.2.1	Develop a public service task force.	International Cooperation
4.3	Expand delivery systems for information transfer to members and the public. *	4.3.1	Conduct member survey for products and services. (6)	Products & Marketing EXCOM in collaboration with GB and GB TC
		4.3.2	Establish public information program. (13)	
		4.3.3	Place "products" on the web. (9)	
		4.3.4	Conduct member survey for products and services. (6)	
4.4	Enable web communications. (10)	4.4.1	Update and maintain conference websites for ease of information retrieval.	Conferences, Continuing Education, and Exhibits Marketing Council/Staff
Goal 5: Implement and Improve the 1999 EWRI Business Plan Governing Board				
OBJECTIVE		ACTION		COMMITTEE/ COUNCIL
5.1	Encourage and reward innovative programs. *	5.1.1	Establish EWRI awards mechanisms. (11)	Awards Committee
5.2	Hold strategic planning and self-evaluation session(s) annually. (1)	5.2.1	Develop EWRI Strategic Plan to chart the course of EWRI's policies, programs, and projects in the new millennium.	Governing Board

TABLE KEY:
Numbers in parenthesis () denote EWRI 2000 Expanded Goals
*Denotes 1996 Goals

History and Heritage of the Irrigation and Drainage Division

Conrad G. Keyes, Jr.[1], *F. ASCE, F. NSPE*, and Kenneth K. Tanji[2], *M. ASCE*

The Irrigation Engineering (IR) Division in ASCE was authorized in June 20, 1922 with 23 members and a five-member Executive Committee elected by members of the Division. The constitution of this Division was fully established by 1925. The Division grew to 645 members and affiliates by 1927. Participants in this Division included engineers from federal and state water-related governmental agencies as well as from irrigation, drainage and reclamation districts, universities, and consulting firms. The IR Division was renamed the Irrigation and Drainage (I&D) Division in 1953. The early action committees organized in the Division included committees on National Reclamation Policy, Cooperation between Federal and State Agencies, Drainage of Irrigated Lands, Duty of Water, and Interstate Water Rights. Over time the Division's activities extended into water conservation, capacity of irrigated lands to pay irrigation costs, sedimentation in reservoirs, irrigation conveyance losses, irrigation and drainage practices in humid areas, drought, reuse of waters for irrigation, on-farm irrigation, and water quality. Many of these topics are still relevant in the 21st Century.

The 1995 ASCE Strategic Plan called for the creation of semi-autonomous Institutes for groups of engineering specialties within ASCE. The Irrigation and Drainage Division and the Hydraulics Division in 1995 were amalgamated into the Water Resources Engineering Division. In 1999, the Environmental and Water Resources Institute (EWRI) was formed from the former Technical Divisions of Environmental (Sanitary in 1922) Engineering, Water Resources

[1] Past President, Environmental & Water Resources Institute, ASCE; Ex-Principal Engineer, US International Boundary & Water Commission; and Prof. Emeritus & Dept. Head, New Mexico State University, Las Cruces, NM 88003 (corresponding author).
[2] Prof. Emeritus, LAWR-Hydrology, Univ. of California, Davis, CA 95616.

Engineering, and Water Resources Planning and Management, as well as the ASCE Water and Environmental Standards Council, with a total membership of about 20,000 in the new Institute of ASCE. The present Irrigation and Drainage Council (IDC) of EWRI of ASCE oversees the activities of Technical Committees such as Water Quality and Drainage, Irrigation Delivery and Drainage Systems, Evapotranspiration in Irrigation and Hydrology, and On-Farm Irrigation. Some of the current task committees include the revisions of ASCE Manuals #57 and #70 (titles provided later), the creation of a I&D Pipeline Manual, and the standardization of Referenced Evapotranspiration.

Prior to the founding of the present ASCE in 1852 and since that time, civil engineers have contributed significantly toward the development of irrigation and drainage in the United States. For example, in 1763 George Washington surveyed the Dismal Swamp area in Virginia and North Carolina for reclamation that led in 1778 to the charter of the Dismal Swamp Canal Company. In 2001, ASCE selected Hoover Dam on the Colorado River, designed by civil engineers and built in the 1930s, as one of the Civil Engineering Monuments of the Millennium. Hoover Dam, at the time of construction, was the highest dam, the costliest water project, and contained the largest power plant. Lake Mead created by Hoover Dam serves as the water supply for about 18 million people and over 0.6 million ha of irrigated land in the U.S.A. and an additional 0.2 million ha in Mexico. The 100th Anniversary of the U.S. Bureau of Reclamation anniversary celebration was held at Hoover Dam on the evening of June 17, 2002 and a Centennial History Symposium at UNLV on June 18-19, 2002.

Civil Engineers in the U.S. have aggressively pursued the heritage established by George Washington and other esteemed engineers related to irrigation and drainage such as Royce J. Tipton (ASCE Society Award in his honor), Harvey O. Banks, Stephen D. Bechtel, Jr., Joseph R. Friedkin, Frank Veihmeyer, Frederick L. Hotes, A. Ivan Johnson, Jack Keller, Finley B. Laverty, James Luthin, William O. Pruitt, Marvin E. Jensen, Jan van Schilfgaarde, Ernest T. Smerdon; to mention a few. Over the past 80 years, civil engineers have participated in the design and construction of major waterworks, irrigation and drainage development, and management of water especially in irrigated agriculture. Technical Committees have been continuously active for decades on such topics as water conveyance and distribution, evapotranspiration, on-farm irrigation, irrigation water requirements, operation and management of irrigation and drainage systems, water conservation, drought, drainage, ground water, water quality, waste reuse, and weather modification. These Technical Committees sponsored numerous specific Task Committees and presented their findings in the Annual I&D Specialty Conferences and published them in ASCE journals, particularly the *Journal of Irrigation and Drainage* and the *Journal of I&D Engineering* of today.

Some of the significant products that the I&D Division has been involved with include ASCE Manuals and Reports on Engineering Practice such as *Hydrology Handbook* (No. 28), *Ground Water Management* (No. 40), *Management, Operation and Maintenance of Irrigation and Drainage Systems* (No. 57), *Evapotranspiration and Irrigation Water Requirements* (No. 70), *Agricultural Salinity Assessment and Management* (No. 71), *Guidelines for Cloud Seeding to Augment Precipitation* (No. 81), *Operation and Maintenance of Ground Water Facilities* (No. 86), and *Urban Subsurface Drainage* (No. 95). Committees that developed under Irrigation & Drainage Division produced ASCE Standards on *Urban Subsurface Drainage* (*Design* is numbered 12-92, *Installation* is numbered 13-93, and *Operation & Maintenance* is numbered 14-93. The recent EWRI/ASCE Standard 34-01, *Standard Guidelines for Artificial Recharge of Ground Water*, was processed through the EWRI Standards Development Council in 1999-2001. The Irrigation and Drainage Division has had an 80-year history of meritorious services and accomplishments to ASCE and to the public.

In this millennium, civil engineers in I&D will be challenged because of increasing loss of agricultural water to urban demands and for fish and wildlife as well as increasing constraints on the quality of irrigation return flows discharged into streams, lakes and ground water basins. Agriculture will need to utilize available freshwater supplies judiciously. Agriculture will have to utilize marginal quality waters in greater quantities than now. New approaches and new inventions will need to be developed to cope with anticipated water problems. Civil engineers will need to be more creative and innovative, and will need to interact more heavily with other professionals (soil and crop sciences, hydrologists and watershed planners, regulators and environmentalists, and economists and policy makers) to successfully provide solutions to water availability and water quality problems.

Summary of the Hydraulics Division- ASCE (1938 – 1988)

Margaret S. Petersen[1], Honorary M. ASCE

INTRODUCTION

Formation of the ASCE Hydraulics Division was authorized by the Board of Direction of the Society in April 1938 as the twelfth technical division. Prior to that time, 28 different committees of the Society had been concerned with water. The petition to form the Division was signed by 200 members of ASCE, with the stated objective:

> To strengthen the position of the civil engineer in general
> and special fields of hydraulic engineering, to sponsor
> papers and encourage discussion of hydraulic
> engineering subjects, to urge civil engineering colleges
> to equip their students with a knowledge equal to that
> imparted in the curriculum of any other type of
> engineering college.

Creation of the Division was largely due to the efforts of Fred C. Scobey, 1937 Chairman of the Society's Committee on Research. Fred Scobey was an irrigation engineer with the U.S. Department of Agriculture who is remembered for his pioneering work on irrigated lands in the West and irrigation canals. He was appointed by the ASCE Board of Direction as the first Chairman of the Division's Executive Committee; other members appointed to the first Executive Committee were: Charles H. Paul, J.C. Stevens, Gerard H. Matthes, and Boris A. Bakhmeteff. The Executive Committee met for the first time in December 1938 and formulated the general program and aims of the Division. Fred Scobey served as Chairman of the Executive Committee from 1939-1942. He was followed by: Boris Bakhmeteff (1943-45), William Hoyt (1946-47), Lorenz G. Straub (1948-49), and Albert S. Fry (1950). Albert S. Fry, who was Chairman of the Executive Committee in 1950, was the person who conceived the idea of the Hydraulics Division holding specialty conferences. Albert Fry was employed by the Morgan Engineering Company for many years prior to joining the Tennessee Valley Authority in1933.   Under him the

[1] 1750 East Rio De La Loma, Tuscon, AZ 85718; phone 520-577-2519; petersen@engr.arizona.com

TVA Hydraulics Laboratory was established, and he directed pioneering work in many aspects of hydraulics and hydrology.

He served as General Chairman of the Division's first Specialty Conference held in Jackson, Mississippi, in November 1950. Such conferences became an annual tradition in the Hydraulics Division, and the idea has since been adopted by other ASCE Divisions. A number of papers at the Jackson conference were presented by Society members who continued to be active in the Division for many years; nine of them later became chairman of the Division's Executive Committee: Morough P. O'Brien, George Hickox, Harold Martin, Carl Kindsvater, Arthur Ippen, Arno T. Lenz, Frederick R. Brown, Rex A. Elder, and Jacob Douma.

A number of awards are made by the Society each year on recommendation of Hydraulics Division. In 1939 Karl E. Hilgard bestowed an endowment for the Karl Emil Hilgard Hydraulic Prize, and the J.C. Stevens Award was established in 1943 by an endowment from John C. Stevens. In recent years members of the Division established two additional awards. The Hydraulic Structures Medal was instituted in 1983 by Fred W. Blaisdell, and the Arid Lands Hydraulic Engineering Award was established in 1986 by Ibrahim M. Elassiouti.

The Hydraulics Division endowed the Hunter Rouse Engineering Lecture in 1979. Purpose of the lecture is twofold: to provide a focal point for the annual conferences, and to honor Dr. Hunter Rouse who has played an important role for more than four decades in the development of engineering hydraulics in this country and internationally. To celebrate the fiftieth anniversary of the founding of the Division, the Executive Committee established a Task Committee charged to develop functions for the 1988 National Conference on Hydraulic Engineering that would recognize and involve ASCE members formerly active in the Hydraulics Division over the last fifty years; The Task Committee organized special sessions and solicited the participation of senior members of the Division to present review papers outlining historical advances in hydraulics and hydrology over the past fifty years and contributions of the Hydraulics Division to the advancement of our profession. Their contributions (were) assembled in this publication in the hope that they will enrich the background of younger Division members, by giving them a sense of the roots of the profession, and also that they will be of interest to those who "were there" for a large part of the Division's past.

As Chairman of the Task Committee on the Fiftieth Anniversary of the Hydraulics Division, I want to thank members of the Task Committee - Frederick R. Brown, Rex A. Elder, and Donald R.F. Harleman - for their invaluable contributions to arrangements for the technical sessions and social events. Last, on behalf of all of us who served on the Task Committee, our most sincere thanks (go) to the speakers and panelists who made the anniversary sessions outstanding events, and to Robert C. MacArthur and Chih Ted Yang who arranged the sediment panel discussion.

> Margaret S. Petersen, Chairman,
> Task Committee on the Fiftieth
> Anniversary of the Hydraulics Division

Historical Development of the
Water Resources Planning and Management Division of the
American Society of Civil Engineers

Bill Cox[1] and Jerry Rogers[2]

The Water Resources Planning and Management Division began its existence as a council, the status it now holds as one of the components of ASCE's Environmental and Water Resources Institute. The predecessor organization was first authorized in 1971 as the Technical Council on Water Resources Planning and Management. This organization was approved as a trial in response to efforts to establish a new division, which had been led primarily by individuals within ASCE's Hydraulics Division and Sanitary (Environmental) Engineering Division. The primary goal of the proposed division was to address a perceived need for a single focal point for water planning and management activities. Such activities had previously been diffused across the various water-related components of ASCE due to their basic importance as a corollary of engineering decision-making related to water.

The initial Council had some of the powers of a division but was subjected to special rules, one of which was the form of the Executive Committee that served as its governing body. As first appointed in 1972, it consisted of one representative fromeach of the water-related divisions. Another restriction relative to a division was the absence of a journal. During this initial period, the Journal of Hydraulics included a special section for publishing papers in the planning and management area. The Council was authorized to hold conferences, and its first general specialty conference was held in 1974 at Cornell University in Ithaca, NY. (The Council's Water Resources Systems Committee held an earlier conference focusing on systems analysis in 1973 in Boulder, CO.)

The Council was officially approved as a Division in 1975. This change meant that the Executive Committee would be selected from within the Division rather than being composed of appointed representatives of the other water-related divisions. The Division also was authorized to publish its own journal. The first issue was published in 1976 as the Journal of the Water Resources Planning and Management Division. The journal name was changed to Journal of Water Resources Planning and Management in 1983.

----

[1] Professor, Department of Civil and Environmental Engineering, Virginia Tech, Blacksburg, VA 24061; phone 540-231-7152; cox@vt.edu.
[2] Professor, Department of Civil and Environmental Engineering, University of Houston, Houston, TX 77204; phone 713-743-4276; jerryrogers@rr.houston.net.

The Division has held an annual specialty conference since the inaugural event in 1974. Following the practice of the first conference, several of these events have occurred on university campuses. Most, however, have been held in major cities representing all the regions of the Nation. The Division traditionally joined the other water-related divisions in the water-forum series held on a five-year cycle, an event similar to the EWRI water congresses initiated in Orlando, FL, in 2001 and scheduled to become an annual event beginning in 2003.

The status of the Division as an independent component of ASCE was evaluated in the early 1990s as part of the proposed merger of divisions into a larger water-focused unit. The associated debate resulted in re-examination of the reasons for formation of the Division. Ultimately, the decision to maintain independent division status was made, and the proposal to join the new Water Resources Engineering Division created in 1993 was rejected. Status was again an issue later in the 1990s as the EWRI proposal was developed. Becoming part of the Institute was seen as a means to address financial and other issues associated with maintaining the status quo while offering the promise of considerable autonomy to continue the programs and traditions of the Division.

Since its early beginnings, the Division's activities have spanned a wide cross section of water planning and management issues. In keeping with the nature of the Division, some of these areas of significant accomplishment have been non-traditional from the engineering perspective. For example, the Division's Water Laws Committee was active in the areas of water allocation law for both the eastern and western United States. This involvement took many forms, with the most notable accomplishment being the initiation of an effort leading to development of model water codes, an activity that has spanned the transition period for creation of EWRI and continues as a major EWRI project. Other focal points of activity have been more traditional such as the systems analysis area. The Water Resources Systems Committee (which was identified above as having held the first conference with WRPM Division involvement) has been an exceptionally productive group throughout the Division's history. The active membership has been responsible for a variety of publications, conference sessions, symposia, and other products.

Singling out individual areas of contribution should not diminish the value of other accomplishments in diverse areas. A variety of other committees made significant contributions, including Application of Emerging Technologies, Economics and Finance, Groundwater Management, International Water Resources Activities, Operations Management, Social and Environmental Objectives, Spill Prevention and Management, Urban Water Resources, Water Policy and Institutions, Water Resources Planning, and Water Supply and Conservation.

The evolution of the WRPM Division has occurred through the personal commitment and leadership of many individuals. Names have not been included in this text, but the history summarized here can be connected to the responsible people through reference to the appendix that lists those who have held governance positions, won Division awards, and made other contributions to Division progress.

Appendix: Founders, National Officers, Leaders, and Selected Award Recipients Technical Council of Water Resources Planning and Management and Water Resources Planning And Management Division American Society of Civil Engineers

Leaders of the initial effort to establish separate Division of Water Resources Planning and Management:
James R. Villemonte, HY (Hydraulic Engineering Division)
Rex A. Elder, HY
Ed Fusick, HY
Walter A. Lyon, SA (Sanitary (Environmental) Engineering Division)
Verne H. Scott, HY

Leaders of the petition for conversion from council to division status:
Victor A. Koelzer
Richard C. Tucker
John S. Gladwell

Executive Committee of the Technical Council on Water Resources Planning and Management:
Victor A. Koelzer, HY, Feb. 1972-1974 (Chair 1972-73)
Maurice L. Albertson, IR (Irrigation and Drainage Division), Feb. 1972-1975 (Chair 1973-74)
Paul W. Eastman, SA, Feb. 1972-1976 (Chair 1974-75)
John W. O'Hara, PO, Feb. 1972-1973
Herbert H. LaVigne, UP (Urban Planning and Development Division)
Eric E. Bottoms, WW (Waterways Division, now WPCOD), Feb. 1972-1973
Herbert S. Reisbol, PO, 1974-1975

Advisor on Water Resources Activities (during TCWRPM):
Leo R. Beard 1972-1973
Gerald T. Orlob 1974-1975

Executive Committee of the Water Resources Planning and Management Division:
Carl H. Gaum, WW, 1974-1976 (Chair 1975-76)
Leo R. Beard, HY, 1974-1977 (Chair 1976-77)
William L. Grecco, UP, 1974-1978 (Chair 1977-78)
Robert E. Fish, 1975-1979 (Chair 1978-79)
Augustine J. Fredrich, 1976-1980 (Chair 1979-80)
Edward Silberman, 1977-1981 (Chair 1980-81)
Jerry R. Rogers, 1978-1982 (Chair 1981-82)
William R. Walker, 1979-1983 (Chair 1982-83)
William K. Johnson, 1980-1984 (Chair 1983-84)
Harold J. Day, 1981-1985 (Chair 1984-85)
Richard M. Males, 1982-1986 (Chair 1985-86)

WRPM Executive Committee, continued
  James R. Hanchey, 1983-1987 (Chair 1986-87)
  Dale D. Meredith, 1984-1988 (Chair 1987-88)
  Glenn M. Johnson, 1985-1989 (Chair 1988-89)
  Darell D. Zimbelman, 1986-1990 (Chair 1989-90)
  Kyle E. Schilling, 1987-1991 (Chair 1990-91)
  Michael A. Ports, 1988-1992 (Chair 1991-92)
  William E. Cox, 1989-1993 (Chair 1992-93)
  Jack M. Mowreader, 1990-1994 (Chair 1993-94)
  Neil S. Grigg, 1991-1995 (Chair 1994-95)
  Jerry L. Anderson, 1992-1996 (Chair 1995-96)
  Katherine A. Hon, 1993-1997
  William A. Macaitis, 1994-1998
  Walter M. Grayman, 1995-1999
  James R. Groves, 1996-2000

Management Group D Representative from WRPMD:
  William C. Ackermann 1972   Augustine J. Fredrich 1981-1985
  Arthur D. Soderberg 1973   Jerry R. Rogers 1985-1990
  Verne H. Scott 1974-1977   Darell D. Zimbelman 1990-1995
  Victor A. Koelzer 1978-1981   Michael A. Ports 1995-2000

National Secretary of theTCWRPM and the WRPMD Executive Committee:
  John J. Gladwell Feb. 1972-1975   Marshall S. Goulding, Jr. 1986-
  J. Ernest Flack 1975-1977   1991
  James D. Goff 1977-1981   Donald M. Phelps 1991-1996
  Terence P. Curran 1981-1984   Richard S. Siegel 1996-
  Jonathan Deason 1984-1986

National News Correspondent of WRPMD:
  Paul F. Ruff 1974-1975   Katherine A. Hon 1985-1987
  Jerry R. Rogers 1976-1978   Harry N. Tuvel 1988-1989
  Glenn M. Johnson 1979-1981   James R. Groves 1990-1994
  Marshall Anthony 1982-1984   Robyn S. Colosimo 1994-

National Editor of the Journal of the WRPMD:
  Edward Silberman   Miguel A. Marino
  Thomas L. Morin   William W.G. Yeh
  Paul F. Ruff   Richard N. Palmer,
  Marshall S. Goulding, Jr.   Jay Lund
  Dale D. Meredith   John Labadie

ASCE Headquarters Contact to WRPMD:

Harry N. Tuvel	Ahmad Habibian
Edward Kippel	Brian Parsons
Marla Berman	Cindy Gold
John Griffin	Autumn Richter
Marla Berman	Wayne Davis

Past Julian Hinds/ National WRPMD Award Recipients:

1975- Victor A. Koelzer	1989- Warren Viessman, Jr.
1976- Harvey O. Banks	1990- William B. Walker
1977- Eugene W. Weber	1991- Theodore M. Schad
1978- Ray K. Linsley	1992- Luna B. Leopold
1979- Carl E. Kindsvater	1993- Augustine J. Fredrich
1980- Dean F. Peterson	1994- William W.G. Yeh
1981- Leo R. Beard	1995- Verne H. Scott
1982- William Whipple, Jr.	1996- Miguel A. Marino
1983- Munson W. Dowd	1997- Jery R. Stedinger
1984- Vincent George Terenzio	1998 – Charles D.D. Howard
1985- Joseph S. Cragwall	1999 – Christine A. Shoemaker
1986- Daniel P. Loucks	2000 – Kyle Schilling
1987- Gerald T. Orlob	2001 – Quentin W. Martin
1988- Robert L. Smith	2002 – George F. Pinder

National Service to the Profession/WRPMD Award:

1986- Augustine J. Fredrich	1995- Kyle E. Schilling
1987- William K. Johnson	1996- Marshall S. Goulding
1988- William R. Walker	1997- Neil S. Grigg
1989- Richard M. Males	1998 – Richard Palmer
1990- Jerry R. Rogers	1999 - Larry Roesner & Jeff
1991- Harold "Jack" Day	Wright
1992- Dale D. Meredith	2000 – Jacques W. Delleur
1993- Conrad G. Keyes, Jr.	2001 – Darrel Zimbelman
1994- Walter A. Lyon	2002 - Hugo Loaiciga

History & Accomplishments of the EWRI/ASCE Codes & Standards Activities

Conrad G. Keyes, Jr.[1], ScD, PE/PS, F. ASCE, F. NSPE

Overview

This paper provides the general policies, operational procedures, and some activities by the Environmental & Water Resources Institute (EWRI) of the ASCE, Standards Development Council (SDC), since October 1, 1999 and its predecessors – the Water & Environmental Standards Council (1995-1999) and the Special Standards Council (1990 – 1994) of ASCE.   The ANSI/ASCE recognized Environmental Standards Committee started in 1977, workshops with funding by EPA began in April 1978, and the Technical Council on Codes and Standards existed within ASCE from 1980 through 1989. The environmental and water resources standards are carefully reviewed consensus documents on design, operation, installation, and/or maintenance for adoption or utilization by the profession. Model Codes or agreements developed by these Councils have to be used by appropriate governmental organizations throughout the world. These procedures and activities intend to provide the history and supplemental information that may be contained in the current editions of the EWRI Operations Handbook, any ASCE Technical Activities Committee (TAC) Handbook, and the 1980-2002 ASCE Official Registers.

Authorization of ASCE CSAC within TAC

The Codes and Standards Activities Committee (CSAC) was authorized by ASCE to be under the administrative authority of the Technical Activities Committee (TAC) in July 1995. CSAC is charged with the overall responsibility of maintaining and administering the Rules and Procedures of the ASCE Standards Program; including those standards organizations within the ASCE Institutes (see Procedures of CSAC, 1999). This

---

[1] Chair, EWRI Standards Development Council and Member, ASCE Codes and Standards Activities Committee, P.O. Box 1499, Mesilla Park, NM 88047, Email: cgkeyesjr@zianet.com

authorization before this point-in-time was provided to Management Group F (MGF) within the same technical activities committee.

Organization of Codes and Standards Program

The Codes and Standards Program of ASCE is comprised of the following groups:

Codes and Standards Activities Committee (CSAC) - has overall administrative authority for the standards program.

Standards Councils - reside within ASCE TAC or the Institutes of ASCE: The councils are generally specific to a single technical specialty area and are responsible for ensuring compliance with the ASCE Rules on Standards, for general marketing of the standards within their discipline, and for stimulating the initiation of new standards activities within their discipline (see the list of the EWRI/ASCE Standards at the end of this paper).

Standards Committees - reside within both ASCE TAC and the Institutes of ASCE: Standards committees are assigned to the standards council or division according to their technical subject. The standards committees and statements of their purpose are listed within the ASCE Official Register each year. A standards committee is the body responsible for the technical content of an ASCE Standard and hence it is the body in which consensus is developed. Publications by standards committees that are not standards must be authorized by their Standards Council and CSAC in advance of initiating the activity and shall be approved through the procedures of the cognizant council/division/Institute and/or as directed by CSAC.

Governance by CSAC

The purpose of CSAC is to provide general oversight and management of the ASCE Standards Program, maintain the ANSI accreditation of the program, and enforce the ASCE Rules on Standards. The duties of CSAC in carrying out this purpose include:

1) Perform all functions as assigned in the ASCE Rules on Standards.

2) Coordinate the administrative and procedural activities of the Society related to the establishment, use, or discontinuance of codes and standards.

3) Recommend to TAC appointment of liaison members to other organizations.

4) Maintain the: *ASCE Rules for Codes and Standards Committee*; *Procedures for the Codes and Standards Activities Committee and Standards Councils*; *Form and*

Style Manual for ASCE Codes and Standards; Strategic Plan for the ASCE Codes and
Standards Activities Program (CSAC, various years); and other similar documents pertaining to the ASCE Codes and Standards Program.

5) Furnish guidance to TAC on policy and general matters relating to codes and standards.

Membership of EWRI SDC Members on CSAC

The members of CSAC shall consist of one ASCE TAC contact (typically a member of the ASCE Board of Direction) and representatives from each of the constituent Councils of CSAC and the Institutes. A Council with one to six Standards Committees shall have one representative on CSAC. A Council with seven to fourteen Standards Committees shall have two representatives on CSAC. A Council with more than fourteen Standards Committees shall have three representatives on CSAC. The EWRI members on CSAC have been Conrad G. Keyes, Jr., ScD, PE/PS, F. ASCE, F. NSPE (served as Vice Chair from 1997-1999) and William C. Boyle. Other members of the "wet" divisions, making up the present EWRI involved in the governance of codes and standards, in the past included A. Ivan Johnson (served as Chair of the historic Special Standards Council for some time), Robert T. Chuck, and Richard Wenberg.

Operations of Standards Council

The purpose of a standards council is to manage its constituent standards committees, enforce the ASCE Rules on Standards, seek opportunities to market the standards which its committees produce, and stimulate the initiation of new ASCE standards activities (see the EWRI Operations Handbook at www.EWRInstitute.org).

The fiscal year of all standards councils begins October 1 and terminates the following September 30. The budget request process for any fiscal year has been different for almost every year since the beginning of the ASCE Codes and Standards process in 1980. Therefore, many of the standards committees have not appreciated the normal standards funding that is conducive to continuously balloting by the committees or for long-term development of new standards.

Membership of the SDC of EWRI

The SDC has consisted of at least three voting members in good standing of either the Society or the EWRI. SDC may have liaison or advisory members from the EWRI Standards Committees and other units of the Institute, as it deems appropriate. The liaison and/or advisory members are normally voting members of the Council.

Members of SDC have been appointed to the Council for a three-year term (CSAC, 1999) and terms shall preferably be staggered. Many of the members are the Chairs or Past Chairs of constituent Standards Committees. SDC Officers are normally elected for two-year terms on the Council. The Standards Council Chair is approved by CSAC may be re-approved.

Meetings of the Standards Council and its Committees

A simple majority constitutes a quorum during meetings. Authorization for meetings shall be obtained in accordance with ASCE Rules for authorization of meetings. The chairs of the units prepare a written agenda for all meetings and send the agenda to the unit members prior to the meeting (the present standards committee Chairs are within the list of EWRI/ASCE Standards at the end of this paper).

At its meetings, the units have:

1) Reviewed the status, plans and accomplishments of each of its constituent units. This may include evaluation of the effectiveness of the units and a review of their purpose and function. The chairs or the liaisons of each Standards Committee shall, at the discretion of the Council, meet with the Standards Council for discussion and the presentation of reports of committee activity.

2) Continuously review and approve the unit's membership.

3) Discussed plans and efforts to initiate new documents or standard activities.

4) Discussed plans for the marketing of their constituent standards including the potential for adoption of the standards by regulatory agencies through which use of the standard may be mandated.

All Standards Councils are encouraged to make recommendations to CSAC for necessary changes to streamline the individual committees and the entire Standards Committees structure including, if necessary, elimination of certain committees or merger with other committees or creation of new committees. The Standards Councils shall also assure inter-Division/Council liaison where a particular topic of study will involve the objectives of more than one committee.

Minutes, Correspondence, and Reports by SDC and Standards Committees

Minutes of all meetings have been prepared and distributed to: (1) CSAC Chairperson, (2) Standards Council members, liaisons and advisory members, (3) Chairs of constituent Standards Committees, (4) ASCE staff for Codes and Standards, (5) Appropriate Institute governing body, and/or (6) Such others as may be appropriate.

Information about each meeting activity suitable for publication in ASCE NEWS or EWRI Currents has been sent to the appropriate News Editor. Copies of significant correspondence on Standards Council or its committees business has been distributed directly to the Standards Council officers with a copy to the appropriate EWRI/ASCE staff. Standards Council's chairs have submitted appropriate Report to their CSAC representatives and copies have been sent to the appropriate staff.

ASCE NEWS is published monthly by the Society. It reserves a column for the standards program. Standards Councils and the appropriate Staff are responsible for providing material to the Editor of ASCE NEWS for publication. Councils' responsibilities have been as follows:

1) To work with staff to ensure publication of all required notices of standards activities and meetings as required by ASCE Rules for Standards Committees.

2) To emphasize news of standards activities. Particularly, highlights from the minutes of committee meetings, photos, and news of Standards Committees developments committees.

3) To maintain an up-to-date list of all ASCE standards being developed within their Standards Council.

4) To publish notices calling for papers for possible presentation at forthcoming national meetings when requested by Standards Committee chairs.

5) To prepare for publication a brief review of each Standards Council meeting based on the minutes.

Other Activities

EWRI Awards as part of the ASCE Codes and Standards Activities Committee (CSAC) Program

The CSAC-2 Award is given to CSAC members and ASCE Board of Direction members appointed to serve on CSAC. It is presented at the expiration of a term on CSAC and has been awarded to Bill Boyle and Conrad Keyes.

The CSAC-3 Award is given to Chairs of Technical Committees who served as advisors and/or actively supported the ASCE Standards Program. One of the outstanding standards committee Chairs to receive this award is Ivan Johnson.

The C/EC (Council/Executive Committee) – 1 Award **is g**iven to the Chair of Council Executive Committee at the expiration of the term of office. If Chair has also served on

CSAC then this award shall recognize both categories. This award has been provided to Bill Boyle, Ivan Johnson, and Conrad Keyes.

The C/EC-3 Award is given to the Chair of an EWRI Standards Committee when the work of the Chair and his committee results in a published ASCE standard or at a reasonable time after the standard has been submitted for publication. Recipients of this award have been Bill Boyle, Ivan Johnson, Conrad Keyes, and Dick Wenberg.

The C/EC-4 Award is given to any member of a standards committee, subcommittee, or task committee who has done exemplary work leading to the publication of an ASCE standard or when the work of the Chair and his committee results in a published ASCE standard. The only EWRI member receiving this award to date is Don Finlayson for his efforts as Secretary and Co-editor of EWRI/ASCE Standard 34-01, *Standard Guidelines for Artificial Recharge of Ground Water*.

International Activities

International standards activities are encouraged per ASCE's International Codes and Standards Policy Statement (PS 365).

Relationship with Other Standards Writing Organizations

As needed, liaison representatives are recommended or elected by CSAC to serve as communicators with other organizations. Staff shall maintain communication with ANSI, NIST, and other significant standards related organizations and, when appropriate, report activities to CSAC (1999).

Marketing Plan for EWRI/ASCE Standards

The primary responsibility for the marketing of specific standards lies with the standards council under which the standard was developed (see the list of EWRI/ASCE Standards at the end of this paper). However, ASCE Publications and the EWRI Publications Council shall recommend activities that encourage marketing of standards. Staff shall provide coordination with the publication agencies of the standards. EWRI/ASCE Staff, CSAC and the Standards Councils shall be active in the pursuit of getting all standards, where appropriate, adopted by the Model Codes and agencies having legal jurisdiction for the standards.

Strategic Plan

CSAC shall develop a strategy for ASCE's Standards Program in accordance with the Society's Strategic Plan. The strategic plan of the EWRI Standards Development Council should include many items as reported in each updated EWRI Strategic Plan (see the

"Initial History and Strategic Planning of the EWRI" by Keyes and Rogers of this publication).

All standards councils, standards committees, and members of standards committees are encouraged to develop short courses and workshops and prepare sessions or papers for presentation at EWRI/ASCE conferences pertaining to their EWRI/ASCE standards.

EWRI STANDARDS TRACKING SYSTEM (Pre-standards in non-ballot process are not included)				
	CONTACT	WHERE PUB CAME FROM	SCHEDULED PUB DATE	INFORMATION
Publication	Contact Name	Committee	MM/DD/YY	Notes
STANDARD PRACTICE FOR THE DESIGN & OPERATION OF HAIL SUPPRESSION PROJECTS	Bruce Boe and George Bomar	ATMOSPHERIC WATER MGT. AND EDITED BY HAIL SUPPRESSION SUBCOMMITTEE	Possibly Completed in Late 2002	THE ASCE PUBLIC BALLOT WAS PROVIDED BY ASCE STANDARDS COORDINATOR AS DIRECTED BY GEORGE BOMAR.IN MAY 2002.
STANDARD PRACTICE FOR THE DESIGN & OPERATION OF PRECIPITATION ENHANCEMENT PROJECTS	Don Griffith and George Bomar	ATMOSPHERIC WATER MGT. AND EDITED BY PRECIPITATION ENHANCEMENT SUBCOMMITTEE	2003	THE THIRD COMMITTEE BALLOT WAS PROVIDED BY ASCE STANDARDS COORDINATOR AS DIRECTED BY GEORGE BOMAR IN MAY 2002

STANDARD PRACTICE FOR THE DESIGN & OPERATION ON FOG DISPERSAL PROJECTS	Tom DeFelice and George Bomar	ATMOSPHERIC WATER MGT. AND EDITED BY FOG DISPERSAL SUBCOMMITTEE	2003	THE FIRST COMMITTEE BALLOT WAS DIRECTED BY BOMAR'S APPROVAL AND WAS SENT OUT BY THE ASCE STANDARDS COORDINATOR IN MARCH 2002. WAITING FOR RESOLUTION REPORT BY DEFELICE.
EWRI/ASCE STANDARD 33-01: COMPREHENSIVE TRANSBOUNDARY INTERNATIONAL WATER QUALITY AGREEMENT	Conrad Keyes, Jr.	BORDER INTERNATIONAL WATER QUALITY - INTERNATIONAL AGREEMENTS SUBCOMMITTEE	Publication Completed	NEXT VERSION WILL BE UNDER THE DIRECTION OF DAENE MCKINNEY.
EWRI/ASCE STANDARD 35-01: *THE GUIDELINES FOR QUALITY ASSURANCE OF INSTALLED FINE-PORE AERATION EQUIPMENT*	Bill Boyle	OXYGEN TRANSFER	Publication Completed	MIKE STENSTROM WILL BE DIRECTING THE NEXT REVISION.
ANSI/ASCE STANDARD 2-91: MEASUREMENT OF *OXYGEN TRANSFER IN CLEAN WATER,* 3rd Edition	Bill Boyle	OXYGEN TRANSFER	2003	ACCORDING TO THE 10/15/01 OT SC MTG MINUTES, THE REVISION WILL BE TAKEN TO PUBLIC BALLOT UNDER THE DIRECTION OF MIKE STENSTROM AND THIS IS ANTICIPATED BY JULY 2002.

ASCE STANDARD 18-96: *STANDARD GUIDELINES FOR IN-PROCESS TRANSFER TESTING*	Bill Boyle	OXYGEN TRANSFER	Publication Completed	MIKE STENSTROM WILL BE DIRECTING THE NEXT REVISION.
STANDARD GUIDELINES FOR THE DESIGN OF URBAN STORMWATER DRAINAGE	Bill Archdeacon and Dick Wenberg	URBAN DRAINAGE	Late 2002	ASCE PUBLIC BALLOT HAS BEEN COMPLETED. WAITING UNTIL ALL THREE STORMWATER DOCUMENTS ARE READY TO PRINT AS ONE SET.
STANDARD GUIDELINES FOR THE INSTALLATION OFURBAN STORMWATER DRAINAGE	Bill Archdeacon and Dick Wenberg	URBAN DRAINAGE	Late 2002	WAITING FOR PRINTING AUTHORIZATION DUE TO FORMATTING
STANDARD GUIDELINES FOR THE OPERATION AND MAINTENANCE FOR URBAN STORMWATER DRAINAGE	Bill Archdeacon and Dick Wenberg	URBAN DRAINAGE	Late 2002	WAITING FOR PRINTING AUTHORIZATION DUE TO MINOR EDITING AND FORMATTING
ANSI/ASCE STANDARD12-92: *STANDARD GUIDELINES FOR THE DESIGN OF URBAN SUBSURFACE DRAINAGE, 2nd Edition*	Dick Wenberg	URBAN DRAINAGE	2003	BILL ARCHDEACON IS DIRECTING THE 6TH LETTER COMMITTEE BALLOT ON THREE ITEMS OF THE REVISED STANDARD
ASCE STANDARD 13-93: *STANDARD GUIDELINES FOR THE INSTALLATION OF URBAN SUBSURFACE DRAINAGE, 2nd Edition*	Dick Wenberg	URBAN DRAINAGE	2003	BILL ARCHDEACON IS DIRECTING THE 6TH LETTER COMMITTEE BALLOT ON ONE ITEM OF THE REVISED STANDARD.

ASCE STANDARD 14-93: STANDARD GUIDELINES FOR OPERATION & MAINTENANCE OF URBAN SUBSURFACE DRAINAGE, 2nd Edition	Dick Wenberg	URBAN DRAINAGE	2003	COMPLETED BY COMMITTEE AND IS AWAITING ASCE PUBLIC BALLOT TO BE PERFORMED AT THE SAME TIME THAT #12 AND #13 ARE READY FOR THE ASCE PUBLIC BALLOTING PROCESS (PUBLISHED IN THE SAME BINDING IN THE PAST).
EWRI/ASCE STANDARD 34-01: STANDARD GUIDELINES FOR ARTIFICIAL RECHARGE OF GROUNDWATER	Ivan Johnson and Don Finlayson	ARTIFICIAL RECHARGE OF GROUND WATER	Publication Completed	DOCUMENT WILL BE REVIEWED FOR REVISION BY 2006.
REGULATED RIPARIAN MODEL WATER CODE	Joe Dellapenna	WATER REGULATORY	2003	COMPLETED 3rd COMMITTEE BALLOT AND RESOLUTION REPORT - NEXT VOTE IN JULY 2002 MAY TAKE THE DOCUMENT TO THE ASCE PUBLIC BALLOT PROCESS.

RE-PUBLICATION OF IMPORTANT DOCUMENTS OF THE ASCE URBAN WATER RESOURCES RESEARCH COUNCIL

By: Members of ASCE's Urban Water Resources Research Council, Represented by Jonathan E. Jones, P.E.[1], Ben Urbonas[2], Larry Roesner[3], Richard Field[4], Shaw Yu[5]

Since its inception in 1963, the American Society of Civil Engineers (ASCE) Urban Water Resources Research Council (UWRRC) has prepared numerous documents that have significantly affected urban water resources management. Many of the leading practitioners in this area have been (or are) members of the UWRRC, and on many publications, these individuals combined their knowledge and insights and produced documents that remain as timely today as when they were first published. Consequently, the present membership of the UWRRC determined that it would be valuable to collect and review the many documents that the UWRRC has prepared over its nearly 40-year history, and to re-publish those that are particularly important. The purpose of this paper is to provide readers with a concise summary of the UWRRC's publications from to 2001.

In 2003, EWRI will publish a book which includes many of the documents listed in the attached table. A UWRRC Task Committee[6] is presently reviewing these documents, and is determining which of them merit re-publication.

This effort is a tribute to the many great contributors to the UWRRC over the past four decades. The list of such people is far too long to publish here, but suffice it to say that many of them have been true icons in the field of water engineering. The authors are privileged to present some of their work, on behalf of the full UWRRC.

---

[1] P.E. and CEO, Wright Water Engineers, Inc., Denver, CO.
[2] P.E. and Chief of Master Planning, Urban Drainage and Flood Control District, Denver, CO.
[3] P.E. and Ph.D., Professor of Civil and Environmental Engineering, Colorado State University, Ft. Collins, Colorado.
[4] Senior Engineer and Project Manager, U.S. Environmental Protection Agency, Edison, NJ.
[5] P.E. and Ph.D., Professor, Department of Civil and Environmental Engineering, University of Virginia, Charlottesville, VA.
[6] The UWRRC thanks two staff members of Wright Water Engineers, Inc. (WWE) in Denver, Colorado who have contributed greatly on this project, Ms. Frankie Lane and Ms. Linda Vandamme.

Initial Summary of Key Publications of the American Society of Civil Engineers Urban Water Resources Research Council From 1965 to 2001

REPORT TITLE	DATE PUBLISHED	AUTHOR(S) or EDITOR(S)	ABSTRACT
Engineering Foundation Research Conference— *Urban Hydrology Research*	August-65		The first Engineering Foundation Research Conference on Urban Hydrology Research was held at Proctor Academy, Andover, New Hampshire, from August 9 to 13, 1965. Since urban hydrology is more than a narrow, technical, engineering subject, the conference included representation of a variety of disciplines. The first day dealt with the broad social, political, financial, legal, and overall planning aspects. The second and third days were devoted to the hydrology and hydraulics of storm drainage, and the fourth day returned to the broader considerations from the federal, state, and local viewpoints. Friday morning summarized the conference.
ASCE Urban Water Resources Research Program TM-2—*Sewage Flow Variations in Individual Homes*	February-66	M.B. McPherson	Very little data is available on flow rates for sewage emanating from individual homes, particularly for short time-intervals on the order of one minute. As an expedient, the data used for this memorandum are all water demand rate. The general concept for separation of combined sewerage systems on which the projet is based involves discharging comminuted or ground sewage from individual buildings and/or building complexes through relatively small pressure tubing laid in existing building connections, and thence into new pressure conduits suspended in existing street sewers. Winter water demands are assumed to represent sewage flows in the absence of sewage flow data. Two samples of

REPORT TITLE	DATE PUBLISHED	AUTHOR(S) or EDITOR(S)	ABSTRACT
			household water demands, recorded at one minute intervals, are used: data from six homes in Maryland for two weeks, and data from two homes in Louisville for four weeks. For each home maximum and minimum 24-hour and 60-, 15-, and 4-minute demands for each day are given. Frequency distributions of 24-hour and 50-minute flows for each sample are compared with each other and with distributions of total flows from groups of three houses and six houses. Based on routing of peak flows from nearly 500 home days of data through various pump storage combinations, a pump capacity of 10 GPM and a total storage capacity of 30 gallons are considered realistic, but not necessarily conservative, estimates for sizing of household sewage storage-pump combinations for ASCE Project applications. Pressure discharge tubing to handle expected flows and head losses for pressure building services would be ¾ to 1¼ inch I.D. Based on studies of peak flows from a six house combination it is tentatively concluded that design flows for gravity street sewers may be suitable for pressure sewer designs in preliminary feasibility studies of the ASCE Project scheme.
ASCE Combined Sewer Separation Project TM-1—*Outline Description of ASCE Project on Separation of Sanitary Sewage from Combined Systems of*	February-66		It is estimated that from 3-5 percent of the annual flow of raw sewage from combined systems of sewerage overflows into lakes, streams, and tidal estuaries and as much as 95 percent of the sewage in combined systems at the time of a storm is emptied into these receiving waters during the progress of the storm. Separation of sanitary sewage from stormwater is considered an important need in the campaign for cleaner waters. A third of the people in the U.S. are estimated to live in communities served by combined sewers.

REPORT TITLE	DATE PUBLISHED	AUTHOR(S) or EDITOR(S)	ABSTRACT
Sewerage			These same people comprise half of the nation's sewered population. It is feared that increasing ubranization will result in a greater rate of increase in organic loads discharged into water courses than can be handled by existing or presently planned corrective efforts. The need for separation of sanitary sewage from combined systems of sewerage and the national scope of the problem of separation is summarized in the appendix to this memorandum.
ASCE Combined Sewer Separation Project TM-2—*Sewage Flow Variations in Individual Homes*	February-67	L.S. Tucker	The nationwide cost for traditional, complete separation of existing combined sewer systems would be very high and diversion of public funds for this purpose would be in competition with projected investment in other public works and services. The basic objectives of the project are to determine the physical feasibility and limitations of the project scheme, or modifications thereto, and to arrive at measures of cost for comparison with the traditional method of separation, for evaluation of investment feasibility.
Major Appliance and Hotpoint Division Appliance Park—Sampling and Analysis of Wastewater from Individual Homes—(Task 2) General Electric	March-67	R.P. Farrell, J.S. Anderson, J.L. Setser	The results of the initial phase of operation of two household wastewater observation stations are described. During a three month period, household wastewater was sampled for analysis, wastewater flow rates were measured, and the behavior of components when handling wastewater under actual conditions of use was observed. Each station included a garbage grinder to macerate incoming sewage solids, a float well and level recorder, a Moyno pump and pump operation time recorder, a check valve on the pump discharge line, and fifty feet of clear plastic discharge tubing. An extensive program of sampling and analysis was carried out to completely characterize the wastewater

REPORT TITLE	DATE PUBLISHED	AUTHOR(S) or EDITOR(S)	ABSTRACT
			from each home. Particulate matter in the wastewater was analyzed over an intensive seven day period to determine its exact nature in terms of particle size, density and microscopic appearance. Water consumption was measured at one gallon intervals at each house and telemetered to recording equipment at the laboratory. A set of fixture tests, during which fixtures were discharged singly and in combinations in preplanned sequences, was run at each station to obtain information on water and sewage flow patterns for fixtures. Daily peak period water demands for 37 days at one station and 48 days at the other are tabulated and frequency curves are shown. The operating history of each station is reviewed.
ASCE Combined Sewer Separation Project TM-3—*Experience with Grinding and Pumping of Sewage from Buildings*	May-67	D.H. Waller	The general concept for separation of combined sewerage systems on which the Project is based involves discharging comminuted or ground sewage from individual buildings and/or building complexes through relatively small pressure tubing laid in existing building connections; and thence through new pressure conduits suspended in existing street sewers. Thirty-six comminutor installations in buildings are described. A description is given of the type of comminutor used in most of these installations, and of typical installation arrangements and recommended maintenance procedures. A curve displays the relative effect of a comminutor and a garbage grinder on sewage particle sizes. A typical garbage grinder is described and use of grinders for toilet wastes is reviewed. Two wet process building-waste pulping systems are described. Two machines that combine the functions of grinding and pumping are

REPORT TITLE	DATE PUBLISHED	AUTHOR(S) or EDITOR(S)	ABSTRACT
			described and the pumping capability of the garbage grinder is dicussed. Practice and experiment in the use of pumps, piping and backflow valves for sewage in buildings is reviewed. An appendix describes the Liljendahl vacuum sewerage system and discusses its relevancy to the ASCE Project scheme.
ASCE Combined Sewer Separation Project TM-4—*Study of Approximate Lengths and Sizes of Combined Sewers in Major Metropolitan Centers*	May-67	Dasel E. Hallmark and John G. Hendrickson, Jr.	The general concept for separation of combined sewerage systems on which the Project is based involves discharging comminuted or ground sewage from individual buildings and/or building complexes through relatively small pressure tubing laid in existing building connections; and thence through new pressure conduits suspended in existing street sewers. In one version of the project scheme, the small pressure tubing would be laid in existing building connections and would then be connected to new pressure conduits suspended in existing street sewers. Entirely different requirements are involved for the installation of pressure conduits in walk-through and non walk-through sewers. Installations in non walk-through sewers would require at least partially remote construction techniques. Direct access would be available without any necessity for excavation only for sewers of a walk-through size. Eleven cities were contacted to obtain information on lengths and sizes of combined sewers. Based on returns from five cities, a tabulation is given of: mileage and percentage of combined sewers greater than 48 inches and percentage of sewers equal to or less than 24 inches. Sizes are based on vertical dimensions of conduits. An average of 72 percent of the sewers are

REPORT TITLE	DATE PUBLISHED	AUTHOR(S) or EDITOR(S)	ABSTRACT
			smaller than 24 inches. Sizes of 54 inches and larger, which are classified as walk-through sewers, account for an average of about 15 percent of the total combined sewer mileage. Results of a storm sewer questionnaire sent out by the American Public Works Association tend to confirm this result for larger population centers. For smaller cities, the percentage is apparently smaller.
ASCE Combined Sewer Separation Project TM-5—*Pressure Tubing Field Investigation*	August-67	L.S. Tucker	The general concept for separation of combined sewerage systems on which the ASCE Project is based involves discharging comminuted or ground sewage from individual buildings and/or building complexes through relatively small pressure tubing. TM-5 relates to special field-trial installations of tubing and conduits for the purpose of determining the nature and extent of practical difficulties that might be encountered in passing various tubing through a building's sewer, in suspending or otherwise attaching a pressure conduit in the street sewer and in making tubing-to-conduit connections. Three methods of installing pressure tubing from houses or small buildings, and of connecting the tubing with street pressure conduits, are described and discussed. For most applications installation and connection of pressure tubing and conduit in trenches, by traditional water distribution methods, is recommended. Field trials for two other methods are described. Tubing was pushed through an 86 foot long 4-inch and 5-inch diameter building lateral, which included three 45° bends, from a specially dug pit at the upstream end into a four foot diameter combined sewer. The forward end of the tubing was guided by a special

REPORT TITLE	DATE PUBLISHED	AUTHOR(S) or EDITOR(S)	ABSTRACT
			leader device. ¾, 1 and 1¼ inch polyethylene tubing could be pushed. Polybutylene and copper tubes could not be pushed because they buckled or crimped. A Kellems grip and swivel on the end of a rope were used to pull tubing from the combined sewer to the upstream pit. . ¾, 1 and 1¼ inch polyethylene and ¾ and 1 inch polybutylene could be pulled. ¼ inch copper tubing could not be pulled because of its stiffness. A connection was made from a ¾ inch polyethylene tube, which had been threaded through the building lateral, to a 2 inch PVC conduit that was temporarily installed in a shallow trench on the building side of the combined sewer. The tubing was brought out of the lateral through a no-hub-wye fitting and connected to the conduit by a corporation cock. The estimated cost of this method would exceed the cost of installing pressure tubing entirely in a new trench between the building and a pressure conduit in a trench dug parallel to the street.
Environmental Engineering Department Central Engineering Laboratories FMC Corporation Report— *Relationship of Sewage Characteristics to Carrying Velocity for Pressure Sewer*	August-67	M.F. Hobbs	Minimum carrying velocities for solid phase matter in smooth plastic 2", 3", 4", 6", and 8" pressure pipes at zero slope were studied for comminuted and uncomminuted raw sewage. The minimum velocity for scouring and the minimum velocity where depositing takes place were essentially the same. Velocities appeared to be independent of the concentration of suspended solids, fixed suspended solids, "sand" concentration, and size distribution of suspended matter and "sand" for the sewages studied. Velocities appeared to be dependent on the size distribution of fixed solids, or more likely the "sand," that acumulated on the bottom of the pipe. Carrying velocities were investigated in an 8" spiral pressure pipe and the results obtained were very erratic. Results

REPORT TITLE	DATE PUBLISHED	AUTHOR(S) or EDITOR(S)	ABSTRACT
			obtained in an 8" plain plastic pipe with open channel flow appeared to agree with results found for 8" pressure pipes and with data reported in the literature. Egg shells that passed through a garbage grinder were carried in the sewage in smooth pressure pipes at lower flow rates than required for scouring the bottom sediments. Average velocities found for the various pipe sizes are given.
Conference Preprint 548—*ASCE Combined Sewer Separation Project Progress*	October-67	M.B. McPherson	This paper is a progress summary for the period from the commencement of the project in February 1966 through August 1967. The general concept for separation of combined sewerage systems on which the ASCE project is based involves pumping comminuted or ground sewage from individual buildings and/or building complexes through relatively small pressure tubing. The tubing would connect to new pressure conduits located outside the buildings. These new, separate sanitary sewage pressure conduits would then discharge into existing interceptors that woud convey the sanitary sewage to treatment works. Stormwater alone would be carried in what were formerly combined sewers. Dr. Gordon M. Fair of Harvard University is the originator of the concept under investigation.
ASCE Combined Sewer Separation Project TM-6—*Hydraulics of a Pressurized Sewerage System and Use of Centrifugal Pump*	November-67	L.S. Tucker	The general concept for separation of combined sewerage systems on which the ASCE Project is based involves discharging comminuted or ground sewage from individual buildings and/or building complexes through relatively small pressure tubing. TM-6 summarizes the general hydraulic requirements for both street sewers and pumping from larger buildings. It provides feasibility studies of hypothetical pressure sewerage systems in the following

REPORT TITLE	DATE PUBLISHED	AUTHOR(S) or EDITOR(S)	ABSTRACT
			combined sewer drainage areas: A 330-acre residential sector in San Francisco, 160-acre residential area in Milwaukee and 60-acre commercial area in Boston. In order to keep each pressure sewer district independent of pressures in the interceptor, a regulating valve is required where a trunk pressure sewer joins an interceptor. The position has been taken that pressure sewers upstream of the interceptor should be kept under positive pressure at all stages of flow. Air relief valves would still be required. Curb pressures would vary from zero to 30 psi. Hydraulic gradients are illustrated for high and low flows and the effects of pressure control devices at the interceptor are illustrated and discussed. For some flat drainage areas sewage pumping would be necessary; a pressure control assembly would be needed immediately upstream, and a surge control valve would be used immediately downstream, of a lift station. For steep drainage areas, pressure control assemblies would be needed to limit maximum pressures. Centrifugal pump characteristics are discussed and information on 32 sewage and solids handling pumps is tablulated. Sewage pump characteristics are such that maximum reasonable limits on discharge rates would be greatly exceeded if variations in total dynamic head were allowed to equal curb pressure variations that are expected in some parts of a pressure sewerage system. Ordinary use of centrifugal pumps in these cases would be precluded. A possible modification of building pumping systems, by throttling pump discharge with a valve controlled to maintain a constant discharge pressure, is discussed. The use of variable speed

REPORT TITLE	DATE PUBLISHED	AUTHOR(S) or EDITOR(S)	ABSTRACT
			drivers is discussed; it is doubtful that their use would be justified.
ASCE Combined Sewer Separation Project TM-7—*Minimum Transport Velocity for Pressurized Sanitary Sewers*	November-67	M.B. McPherson, L.S. Tucker and M.F. Hobbs	The general concept for separation of combined sewerage systems on which the ASCE Project is based involves discharging comminuted or ground sewage from individual buildings and/or building complexes through relatively small pressure tubing. The required size of a street pressure conduit is governed by two considerations: 1) the magnitude of the lowest flow rate at which transport of all solids is desired, and 2) the minimum mean flow velocity needed to move all solids. The major part of this memorandum is a summary interpretation which provides general criteria applicable to the design of pressure conduits and which sets forth tentative findings for open channel flow. The memorandum also defines design criteria used by project staff for the study of three hypothetical pressurized sanitary sewer systems. In sewage transport velocity tests, in 2-inch through 8-inch clear plastic pipes, sewage sand content was determined. Results are expressed in terms of an empirical relationship developed by Larsen, and are correlated with results of work by Craven and Ambrose. Scouring in pressure pipes requires somewhat higher velocities than depositing; no obvious difference was evident for four open channel tests. Neither comminution nor grinding particularly changed sewage sand size distribution. Size distributions of sewage sands and sands used by Ambrose and Craven are compared. Within the range of experimental data, the effect of particle size has little effect on transport requirements. A reasonably precise prediction of

REPORT TITLE	DATE PUBLISHED	AUTHOR(S) or EDITOR(S)	ABSTRACT
			minimum scouring velocities is dependent upon expected total sand concentration. A linear multiple regression analysis of the sewage flow data indicates that the following variables, in order of decreasing importance, affect scouring velocities: diameter, sand concentration, sand specific gravity, median sand sieve size, and ratio of sieve size passing 90 percent to median sieve size. For preliminary hydraulic design of hypothetical pressure sewerage systems for evaluation of physical feasibility and cost attractiveness, the expression $V = \sqrt{D_{/2}}$ (V is ft/sec, D is internal conduit diameter in inches) has been chosen. This relation satisfies the bulk of the data and provides a degree of conservatism consistent with the level of precision of design flow rates.
ASCE Combined Sewer Separation Project TM-8—*Domestic Sewage Flow Criteria for Evaluation of Application of Project Scheme to Actual Combined Sewer Drainage Areas*	November-67	M.B. McPherson	The general concept for separation of combined sewerage systems on which the ASCE Project is based involves discharging comminuted or ground sewage from individual buildings and/or building complexes through relatively small pressure tubing. The basic objectives of the project are to determine the physical feasibility and limitations of the project's scheme, or modifications thereto, and to arrive at measures of cost for comparison with the traditional method of separation, for evaluation of investment feasibility. The nationwide cost for traditional, complete separation of existing combined sewer systems would be very high and diversion of public funds for this purpose would be in competition with projected investment in other public works and services. Sizing of pressure sewers must be based on attainment of minimum necessary scouring

REPORT TITLE	DATE PUBLISHED	AUTHOR(S) or EDITOR(S)	ABSTRACT
			velocities at an acceptable level of frequency, such as for at least one hour each day of the year. Deposition of solids part of the time must be accepted to preclude ultra-conservative design. At the other extreme, peak sewage rates will give rise to maximum hydraulic gradients, and these in turn affect pressure requirements within the system, for booster pumping and/or pressure control. Residential sewage flow criteria are developed for use in design of pressurized sanitary sewers that form part of the ASCE Project scheme. In a typical combined sewer area, data on annual domestic water demands is the most that can be expected to be available. On the basis of a study of winter water demand data, it is concluded that refinements in projections to account for differences in demand related to numbers of dwelling occupants would be unrealistic, that estimates of mean annual domestic demand based on data from other communities would contain considerable inherent errors, and that projection of observed demands for the drainage area under study to the end of the design period is a preferred basis of design. Winter water demand data for groups of houses of varying sizes is presented in the form of ratios—to the annual average demand rates—of rates on the minimum day, maximum day, maximum hour, and maximum hour on the minimum recorded day. Data for California and the northeastern United States are presented separately, and are plotted to show how ratios to the annual average are related to numbers of dwelling units. For each region design curves are selected to represent the variation, with numbers of dwelling units, of flows for the minimum 24 hours, for the

REPORT TITLE	DATE PUBLISHED	AUTHOR(S) or EDITOR(S)	ABSTRACT
			peak hour of the minimum day, and for the maximum peak hour of any day—all presented as ratios to the annual average rate.
FWPCA Contract No. 14-12-29, Tasks 7 and 9—*Report to ASCE Combined Sewer Separation Project on FWPCA Contract No. 14-12-29, Tasks 7 and 9*	December-67	Robert N. Bowen, John G. Havens	Assistance was provided in connection with special field trial installations of flexible tubing in building sewers. Materials were proposed for pushing or pulling through a building sewer and a methodology and necessary attachments and tools were recommended. Experience gained from the field trials—which are considered to have been very successful—is reviewed and used as a basis for recommendations about pressure sewerage system design and operation. Polyethylene and polybutylene tubing were recommended for use inside building sewers; copper tubing could be used in open trenches. Scratching or scoring was not a serious problem in fishing tubing through a building sewer. The recommended methodology for installation was sound. A saddle type of connection is recommended for connection of pressure tubing to street pressure conduits. Experience with plowing of pressure pipe is reviewed; use of this method where paving or buried utilities are not problems should be investigated. Reference is made to standard practice for trench installations, street crossings and thrust blocking. Two methods of cleaning house pressure tubing are proposed. Six possible layouts of pressure conduits are discussed in terms of operation and maintenance of each. All six arrangements provide for routine rerouting of flow by use of a dual configuration for all conduits.
ASCE Combined Sewer Separation Project TM-	January-68	D.H. Waller	The general concept for separation of combined sewerage systems on which the ASCE Project is based involves discharging comminuted or ground sewage from individual buildings and/or building

REPORT TITLE	DATE PUBLISHED	AUTHOR(S) or EDITOR(S)	ABSTRACT
9—*Peak Flows of Sewage from Individual Houses*			complexes through relatively small pressure tubing. The specific purposes of this memorandum are to present recent data on sewage flows at two household observation stations, to use this data to estimate upper limits on pump and storage capacities for individual houses, and to examine the relationship between peak rates of sewage flow and corresponding water demand rates. The memo: 1. Reviews literature relating to flows of sewage from individual houses. 2. Describes collection and analysis of sewage flow data and related water demand data. 3. Describes the results of fixture discharge tests that were conducted at two household stations. 4. Uses the information above to develop hydrographs from maximum periods of sewage discharge. 5. Presents information on peak sewage flows for a 14-day period at a representative house and examines the relationship between peak sewage flows and related peak water demands. Sewage flows and corresponding water demands were measured at two household observation stations. Water and wastewater flows from individual fixtues and appliances were also measured. The resulting data is presented in detail and is used to estimate upper limits on pump and storage capacities for individual houses and to examine the relationship between peak rates of sewage flow and corresponding water demand rates. Literature relating to flows of sewage from individual houses is reviewed. For individual fixtures,

REPORT TITLE	DATE PUBLISHED	AUTHOR(S) or EDITOR(S)	ABSTRACT
			combinations of rate, duration and frequency of discharge that will produce maximum hydraulic loading conditions are selected. The pattern of discharge of two fixtures discharged simultaneously is not significantly different from the pattern obtained by adding discharges of individual fixtures. Single fixture hydrographs are combined to produce synthetic hydrographs of peak period sewage discharge from houses. The hydrographs are used to determine pump-storage combinations for use in individual houses. Mass curves and frequency distributions, for peak sewage flows and parallel water demands at one house for a 14-day period, are presented. Frequency distributions of water demand data are compared with distributions of four other samples of data and are found to be consistent. Relationships between individual house sewage flows and causative water demands, the effects of numbers of household occupants on mean demand, and the relationship of peak demands to mean demands and to numbers of fixtures and appliances, are examined.
ASCE Combined Sewer Separation Project TM-10—*An Examination of the Benefits and Disadvantages with Respect to the Disposal of Solid Wastes*	February-68	D.H. Waller	The general concept on which the ASCE Combined Sewer Separation Project is based involves the discharge of comminuted or ground sewage from buildings and/or building complexes, via relatively small pressure tubing, into new pressure sanitary sewers. The new pressure sanitary sewers would discharge into existing interceptors that would convey the sanitary sewage to treatment works; stormwater would be conveyed in what were formerly combined sewers. The ASCE Project concept envisions the installation of new systems of pressure sanitary sewers as a method of diverting sanitary sewage from existing combined

REPORT TITLE	DATE PUBLISHED	AUTHOR(S) or EDITOR(S)	ABSTRACT
			sewers. The possibility of obtaining synergistic benefits by adapting the Project scheme to the disposal of solid wastes was recognized early in the Project program. The purpose of this memorandum is to examine considerations that are important to an evaluation of the feasibility and benefits of adapting the Project scheme to solid wastes disposal. Most of these considerations apply to any system of sanitary sewers. Considerations common to both open channel and pressure sanitary sewers are discussed first, followed by an examination of considerations peculiar to adaptation of the Project scheme to the disposal of solid wastes. Important considerations in an evaluation of the feasibility and benefits of adapting any sewerage system to solid wastes disposal are discussed first: the extra solids load that community refuse could add to a sewage disposal system; velocities required to move solid wastes and the effect of flow variations on sewer velocities; solid wastes separation practices and attitudes toward separation of household refuse; the need for ringding, and considerations involved in the development of a household refuse grinding device; the effects of solid wastes on sewage treatment processes and costs and benefits involved in evaluation of alternative systems for disposal of sewage and solid wastes. Considerations that are peculiar to the ASCE Project scheme are: the possibility of adapting building sewage grinder-storage-pump units to grinding and discharge of solid wastes; the need to discharge solid wastes into the system under pressure; reduced clearances in the smaller pipes of a pressure system; reduced

REPORT TITLE	DATE PUBLISHED	AUTHOR(S) or EDITOR(S)	ABSTRACT
			water requirements, when compared with an open-channel system, to obtain the same velocity by flushing, and the possibility of obtaining higher velocities within the limitations of design capacity; and the possibility of greater solids deposition at low flows. Appendices provide information on: composition and characteristics of solid wastes; pertinent solid wastes research and development; results of research on transport and treatment of solid wastes in sewage disposal systems; and four systems of building solid wastes disposal.
ASCE Combined Sewer Separation Project TM-11—*Control Techniques for Pressurized Sewerage Systems*	March-68	James R. Daneker & William H. Frazel	The general concept for separation of combined sewerage systems on which the ASCE Project is based involves discharging comminuted or ground sewage from individual buildings and/or building complexes through relatively small pressure tubing. The tubing would connect to new transmission conduits located outside the buildings. The new separate sanitary sewage street pressure conduits would then discharge into existing interceptors that would convey the sanitary sewage to treatment works. Stormwater alone would be carried in what were formerly combined sewers.
			Instrumentation and control of a pressurized sewerage system is well within current technology, and special designs are foreseen that can approach zero maintenance. A rubber-seated butterfly valve is recommended as the final control element. A diaphram pressure element would avoid problems due to sewage solids. A venturi type flow element or a magnetic flow meter could be used, and where accurate measurement is not a prime requirement, it is likely that a simplified version of a mass flow meter could be

REPORT TITLE	DATE PUBLISHED	AUTHOR(S) or EDITOR(S)	ABSTRACT
			worked out. For control of pressure in response to flow changes, a transducer will be required to generate some characterized signal that will be the set point for a pressure controller. The transducer should incorporate a cam that can be cut in the field. System No. 1 maintains a fixed pressure upstream of the control element by modulation of the valve to correct or reduce any deviation of measured pressure from a selected set point. System No. 2 modulates the valve to maintain a specific upstream pressure corresponding to every rate of flow measured at the flow element. System No. 3 would control the start-stop sequence of a booster or lift station centrifugal pump to permit starting without surge and maintain a constant discharge pressure. For a booster station this system would vary pump speed in response to suction pressure. For nearly fool-proof fail-safe control an all hydraulic control system is recommended in preference to pneumatic, hydro-pneumatic, or electronic systems.
ASCE Combined Sewer Separation Project TM-3A— *Experience with Grinding and Pumping of Sewage from Buildings (Concluded)*	March-68	D.H. Waller	The basic objectives of the Project are to determine the physical feasibility and limitations of the Project scheme or modifications thereto, and to arrive at measures of cost for comparison with the traditional method of separation for evaluation of investment feasibility. Results of monitoring the operation of 36 comminutor installations that serve individual buildings are reported. The monitoring program, which covered periods of up to 16 months, is described. Descriptions and prior operating histories of the installations are included. Twenty-four of the installations include sewage pumps following the comminutors. Discharge mains, three to six inches in diameter, vary from less than 50 feet to

REPORT TITLE	DATE PUBLISHED	AUTHOR(S) or EDITOR(S)	ABSTRACT
			more than 6,500 feet in length. Frequency of attention and maintenance is recorded and compared with manufacturers' recommendations. Twenty-five of the machines were inspected at least five times each week. Eighteen machines did not jam during the record period; average frequency of jamming for all machines was about once in six months. There appears to be no significant relationship between frequency of jamming and frequency of cleaning. Replacement or sharpening of cutting elements is more likely to be carried out as a result of damage to parts than a specific effort to maintain sharp cutting edges. Cutting elements were replaced in three machines. Rag fibers, which after passing through a comminutor tend to "rope" in a turbulent environment, and thin pieces of rubber, that are not shredded by the machine, are potential sources of trouble where projections are available where the might accumulate. An appendix contains a summary description of a system, developed at Pennsylvania State University, for conservation of water in residences by recycling.
Summary Report— *Metropolitan Water Resources Research Agenda Committee Meeting*	April-68		The Office of Water Resources Research has been asked by the Federal Council for Science and Technology—Committee on Water Resources Research to develop a recommended national program on metropolitan water resources research. In response to that assignment, an Agenda Committee was organized to assist in defining the scope of a program of research on metropolitan water problems, and in formulating the approach to be followed. The committee met on April 17 and 18, 1968, at the Shoreham Hotel in Washington, DC.

REPORT TITLE	DATE PUBLISHED	AUTHOR(S) or EDITOR(S)	ABSTRACT
			The job of the Agenda Committee was to outline the scope and character of the research effort. The task of specifying the research needed will be undertaken subsequencly by a series of technical panels working under the guidance and direction of a Steering Committee. The objective is to have ready by November 1968, a report containing the recommendations for research which will be necessary to cope with the water resource problems of metropolitan areas. That final report will also include estimates of the funding, scheduling, and manpower requirements to accomplish the proposed research. This report summarizes the presentation and discussion of the participants at the initial Agenda Committee meeting. It also presents the Committee's judgments about the action to be taken to develop the proposed research program by November 1968.
General Electric— *Long-Term Operation of Wastewater Observation Stations (Task 2)*	April-68	R.P. Farrell	In the terminal phase of operation of two houschold wastewater observation stations, the stations were operated for seven months during which the principal object was collection of usage experience. Each station contained a grinder to macerate incoming wastewater solids, a wet-well, a Moyno pump and pump operation time recorder, a discharge ball-check valve, and fifty feet of clear PVC discharge tubing. The two garbage grinders were never a source of difficulty. The principal problem with pumps was wear of non-metallic rotors. The ¾ inch discharge tubing in the last month of operation of one station is attributed to a thick coating of anaerobic slime on the interior walls, which is attributed to low inflow rates, small

REPORT TITLE	DATE PUBLISHED	AUTHOR(S) or EDITOR(S)	ABSTRACT
			occupancy, and extended periods of disuse. Tubing at the other station was essentially clean except for intermittent discolorations that were attributed to migration of the plasticizer out of the tubing. Operating problems that occurred only once or twice are listed and explained. Results of fixture flow tests, and information on overflows from the station wet-wells, was obtained to supplement results of earlier studies.
ASCE Combined Sewer Separation Project TM-12—Non-Mechanical Considerations Involved in Implementing Pressurized Sewerage Systems	May-68	D.H. Waller	The general concept on which the ASCE Combined Sewer Separation Project is based involves the discharge of comminuted or ground sewage from buildings and/or building complexes, via relatively small pressure tubing, into new pressure sanitary sewers. The new pressure sanitary sewers would discharge into existing interceptors that would convey the sanitary sewage to treatment works; stormwater would be conveyed in what were formerly combined sewers. The Project scheme was conceived by Dr. Gordon M. Fair of Harvard University. The basic objectives of the Project are to determine the physical feasibility and limitations of the Project scheme or modifications thereto, and to arrive at measures of cost for comparison with the traditional method of separation for evaluation of investment feasibility. The proposal, implicit in the ASCE Project concept, to install a grinder-storage-pump unit in every building, raises questions regarding: allocation of costs associated with the units; responsibility for the results of malfunction of the units; arrangements for service of the units; and willingness of owners to accept the units in their buildings. Twenty-five householders in

REPORT TITLE	DATE PUBLISHED	AUTHOR(S) or EDITOR(S)	ABSTRACT
			Radcliff, Kentucky, whose houses are served by sewage pumping units, were interviewed to obtain opinions about features of the units that appeared to represent potential sources of nuisance, inconvenience, or other liabilities. Also interviewed were the superintendent of the utility operating the Radcliff sewerage system, owners of five houses in Louisville, Kentucky, at which sewage sampling stations were located, and three consulting engineering firms who have considered schemes involving the installation of sewage pumping equipment on private properties. Opinions and practices reported reflect the view that sewage pumping equipment placed on private property as a part of a public project should be purchased, installed, and serviced at public expense. Even if they were relieved of the direct costs, indirect costs, real and intangible, might discourage property owners from accepting the units. Acceptance of the units could be increased by publicly operated service organizations, community acceptance of responsibility for the results of flooding due to unit malfunction, and reduction of sewer service charges to offset costs associated with the units.
ASCE Combined Sewer Separation Project TM-13—*Special Requirements for a Full Scale Field Demonstration of the ASCE Combined Sewer*	June-68	D.H. Waller	The general concept on which the ASCE Combined Sewer Separation Project is based involves the discharge of comminuted or ground sewage from buildings and/or building complexes, via relatively small pressure tubing, into new pressure sanitary sewers. The new pressure sanitary sewers would discharge into existing interceptors that would convey the sanitary sewage to treatment works; stormwater would be conveyed in what were formerly combined sewers.

REPORT TITLE	DATE PUBLISHED	AUTHOR(S) or EDITOR(S)	ABSTRACT
Separation Project Scheme			This memorandum is essentially an annotated checklist of matters that should be taken into account in planning a field demonstration of the overall project scheme, described in the paragraph above. Its purpose is to delineate problems that might be encountered in planning, constructing and operating a pressure sewerage system, to detail unknowns and effects currently subject to question, and to suggest minimum required or desirable field measurements, observations and sampling for an adequate demonstration of the scheme. Some parts of this memorandum are illustrated by reference to a hypothetical installation in a 160 acre combined sewer drainage area, presently containing about 500 buildings, in Milwaukee, WI. To provide background information, an abbreviated study has been made of problems and costs of combined sewer separation on private property in the ten largest cities that have combined sewerage systems.
Report on the Second Conference on Urban Water Resources Research— *Water and Metropolitan Man*	August-68	William Whipple, Jr., et al.	Although all aspects of urban hydrology research continue to be of vital importance to the nation, the need to refine identifications and understandings of inter-disciplinary interfaces is still of paramount importance. Such refinement is fundamental to effective implementation of systematic (systems) approaches to metropolitan water resources problems. This conference, therefore, had as its objective the fostering of an awareness of the metropolitan establishments of the problems by involving sociologists, political scientists, administrators, planners, engineers, etc. The conference examined the role of water in the metropolitan establishment and explored the optimum

REPORT TITLE	DATE PUBLISHED	AUTHOR(S) or EDITOR(S)	ABSTRACT
			involvement of the various disciplines and pursuits. It is felt that this Conference record indicates the high interest of the urban water engineer in the social problems and impacts of metropolitan water resources and a determination on his part to welcome the necessary collaboration of the sociologist, economist, planner, administrator, lawyer, political scientist, the systems designer, and the Federal Urban Research official in the clearly necessary team approach required to most satisfactorily solve urban water resources problems. In turn the non-engineering participants in the Conference clearly evidenced their high interest in and desire to participate in the multi-disciplinary approach to metropolitan water problems.
A Study of ASCE Urban Water Resources Research Council (UWRRC) - ASCE TM-1— *Northwood Gaging Installation, Baltimore - Instrumentation and Data*	August-68	L.S. Tucker	This technical memorandum is a detailed report on a small urban drainage area called Northwood located in Baltimore, Maryland. Northwood was one of several urban drainage areas instrumented by the Storm Drainage Research Project, Johns Hopkins University, under the direction of Dr. J.C. Geyer, Head, Department of Environmental Engineering Science. This report was made possible through the cooperation of Dr. Geyer. Substantial assistance was given to the writer by Mr. D. Horn, last Project Engineer of the former Storm Drainage Research Project.
A Study of ASCE Urban Water Resources Research Council (UWRRC) - ASCE TM-2— *Oakdale*	August-68	L.S. Tucker	This technical memorandum is a detailed report on a small urban drainage area located in Chicago, Illinois. The drainage area was instrumented by the Chicago Department of Public Works, Bureau of Engineering, under the direction of Mr. C.J. Keifer, Chief Water and Sewer Design Engineer for the Bureau of Engineering. This report was made possible through the

REPORT TITLE	DATE PUBLISHED	AUTHOR(S) or EDITOR(S)	ABSTRACT
Gaging Installation, Chicago-Instrumentation and Data			cooperation of Mr. Keifer. Substantial assistance was given to the writer by Mr. D. Westfall of Mr. Keifer's staff.
A Study of ASCE Urban Water Resources Research Council (UWRRC) - ASCE TM-3— *Response Characteristics of Urban Water Resource Data Systems*	August-68	J.C. Schaake, Jr.	This report is concerned with the collection of data. The need for these data is apparent to many who have worked with urban water resources, but different needs seem to exist for different reasons. This creates a serious problem because the financial resources to collect these data are limited.
A Study of ASCE Urban Water Resources Research Council (UWRRC) - ASCE TM-4— *A Critical Review of Methods of Measuring Discharge Within a Sewer Pipe*	September-68	H.G. Wenzel, Jr.	This report consists of a review and evaluation of existing devices and methods which might be used for discharge measurements in sewers, keeping in mind the data requirements and physical limitations imposed by their use in an urban pilot area. In addition, rating curves for a suggested critical flow device are presented and recommendations are made for future research which, in the author's opinion, is required.
ASCE Combined Sewer Separation Project— *Report on Pressure Sewerage System, Summer Street*	September-68	ASCE Project Staff and Camp, Dresser & McKee	The research upon which this report is based was performed pursuant to Contract No. 14-12-29, with the Federal Water Pollution Control Administration, Department of the Interior. As part of this contract, the ASCE is studying methods of separating combined sewerage systems in three major cities of the United States. This report is the result of a study of the 50-acre Summer Street Separation Study

REPORT TITLE	DATE PUBLISHED	AUTHOR(S) or EDITOR(S)	ABSTRACT
Separation Study Area, Boston MA			Area of Boston, MA. The main objective of this report is to evaluate and compare the conventional gravity separation method (the construction of a new gravity sanitary sewerage system) and pressure methods (the construction of new pressure sanitary sewerage systems consisting of pumping facilities and force mains).
Report prepared for ASCE by Brown and Caldwell Consulting Engineers San Francisco, CA— *Separation of Combined Wastewater and Storm Drainage Systems—San Francisco Study Area*	September-68	Brown and Caldwell Consulting Engineers	This report has been prepared for the ASCE. The research upon which this report is based was performed pursuant to Contract No. 14-12-29 with the Federal Water Pollution Control Administration, United States Department of the Interior. A detailed description of the research study has been prepared by ASCE and published as Technical Memorandum No. 1. Briefly stated, the objectives of the ASCE study are (1) to determine the feasibility of grinding or comminuting sanitary sewage from individual buildings or groups of buildings and conveying it to existing interceptors through a system of pressure conduits, and (2) to compare and evaluate costs and other factors for the pressure system and those for more traditional methods of separation. Work under item (1) also includes development of equipment and construction techniques applicable to the pressure system.
Report— *Systematic Study and Development of Long-Range Programs of Urban Water Resources Research*	September-68	ASCE	ASCE is assisting in outlining, developing and initiating a coordinated national program of urban water resources research. Deliberate and systematic study of urban water problems on a comprehensive basis had long been neglected. The objective of this research is to provide guidelines for initiating and expanding a program of long-range studies in urban water problems. This report covers the first year of work. Urban water resources planning and supporting research are at levels well below

REPORT TITLE	DATE PUBLISHED	AUTHOR(S) or EDITOR(S)	ABSTRACT
			those for river basins and large regional complexes. Deliberate and systematic study of urban water problems from an overall point of view had long been neglected. This report covers the first year accomplishments of an ASCE project, supported by OWRR, to assist in outlining, developing and initiating a coordinated national program of urban water resources research. The project will also have significant international benefits.
Report— *Pressure Sewerage System— Summer Street Separation Study Area*	September-68	ASCE Project Staff and Camp, Dresser & McKee Consulting Engineers, Boston, MA	The research upon which this report is based was performed pursuant to Contract No. 14-12-29, with the Federal Water Pollution Control Administration, Department of the Interior. As part of this contract, the ASCE is studying methods of separating combined sewerage systems in three major cities of the United States. This report is a result of a study of the 50-acre Summer Street Separation Study Area of Boston, MA. The main objective of this report is to evaluate and compare the conventional gravity separation method (the construction of a new gravity sanitary sewerage system) and pressure methods (the construction of new pressure sanitary sewerage systems consisting of pumping facilities and force mains). CDM, in accordance with its contract with the ASCE, has conducted engineering investigations in two phases as follows: 1. Phase I—A study of revisions necessary on a typical building to in-house plumbing for separation of stormwater and sanitary sewage both for gravity and pressure separation and including a description of the physical problems involved and construction cost

REPORT TITLE	DATE PUBLISHED	AUTHOR(S) or EDITOR(S)	ABSTRACT
			estimates for these revisions. 2. Phase II—A review of the layouts for the system of pressure separation of the entire study area prepared by the Project Staff and the preparation of construction cost estimates for the systems.
A Study of ASCE Urban Water Resources Research Council (UWRRC) - ASCE TM-5—*The Nature of Changes in Urban Watersheds and Their Importance in the Decades Ahead*	December-68	M.B. McPherson	The basic purpose of the ASCE UWRRC is to help establish coordinated long-range research in urban water resources on a national scale. The program currently consists of two major projects: "Research and Analysis of National Basic Information Needs in Urban Hydrology," sponsored by the Geological Survey; and a "Systematic Study and Development of Long-Range Programs of Urban Water Resources Research," sponsored by the Office of Water Resources Research.
Report on Milwaukee Study Area—*ASCE Combined Sewer Separation Project*	December-68	ASCE Project Staff and Greeley and Hansen Consulting Engineers, Chicago, IL	This report has been prepared for the ASCE. The research upon which this report is based was perfromed pursuant to Contract No. 14-12-29 with the Federal Water Pollution Control Administration, Department of the Interior. This report is part of an overall research study being conducted by the ASCE to determine the feasibility of separating combined sewerage systems by using a system of pressure conduits to convey sanitary sewage from individual structures to an existing interceptor. A study area in Milwaukee, WI has been picked by the ASCE project staff. The report includes a detailed description of plumbing changes required to separate sanitary wastes from roof drains, and the

REPORT TITLE	DATE PUBLISHED	AUTHOR(S) or EDITOR(S)	ABSTRACT
			work required to install a grinder-storage-pump unit in each building capable of discharging comminuted sewage to a pressure collection system located in the public right-of-way. Considerable effort was made in determining the plumbing changes required to accomplish separation of wastes within buildings, as little work has been done in this part of combined sewage separation. Two alternative layouts of a pressure sewer system have been submitted by L.S. Tucker, Deputy Project Director of the ASCE project staff. A cost estimate of the pressure layouts has been made and compared to the cost of accomplishing in-house separation and area collection of wastes by the conventional gravity sewer system.
Report—General Electirc Research and Development Center—*Advanced Development of Household Pump-Storage-Grinder Unit*	December-68	R.P. Farrell	In a modern metropolitan sewerage system, sanitary wastewater is collected in separate sanitary sewers and thereby conveyed to treatment works. Stormwater is collected in separate storm sewers and discharged into watercourses. Most of the older cities in the United States have combined sewerage systems, in whole or in part. Both sanitary sewage and stormwater are carried in combined sewers. During dry-weather conditions the entire flow is sanitary sewage, which can be fully diverted to a treatment plant. With significant precipitation, the capacity of diversion structures is exceeded and a mixture of sanitary sewage and stormwater overflows the diversion barriers and discharges into receiving watercourses. Serious concern has been expressed in recent years over the pollution of waterways caused by overflows from

REPORT TITLE	DATE PUBLISHED	AUTHOR(S) or EDITOR(S)	ABSTRACT
			combined sewers, and Congress has appropriated funds for research on alternative methods of reducing this pollution. A late 1967 estimate of the national construction cost for the most direct method, diverting sanitary sewage from individual buildings into new sanitary sewers, is $48 billion.
A Study of ASCE Urban Water Resources Research Council (UWRRC) - ASCE TM-6— *Some Notes on the Rational Method of Storm Drain Design*	January-69	M.B. McPherson	The "rational method" is the prevailing technique currently employed in determining the capacity of storm sewers in design. Improved techniques will certainly evolve from projected plans to secure data on rainfall-runoff-quality for a variety of catchment areas and to develop analytical models for process replication. The purpose of this technical memorandum is to review some of the more salient features and attributes of the rational method. It is hoped that this presentation will indicate why the rational method is no longer adequate.
A Study of ASCE Urban Water Resources Research Council (UWRRC) - ASCE TM-7— *A Study of the Expenditures for Urban Water Services*	February-69	LeRoy H. Clem	The ASCE has undertaken, for the Office of Water Resources Research, Department of the Interior, a contract to assist in outlining and developing a coordinated national program of urban water resources research. Present replacement value of urban facilities for public water, wastewater, and storm drainage totals $110 billion and is increasing at more than $7 billion per year.
A Study of ASCE Urban Water Resources Research Council (UWRRC) - ASCE TM-10—*Sewered*	March-69	L.S. Tucker	The USGS Project on "Analysis of National Basic Information Needs in Urban Hydrology" is primarily aimed at improvement in design and management of urban storm drainage with a predominant emphasis on sewered catchments. Data are virtually non-existent that are suitable for the analysis of rainfall runoff-quality relationships, and that have been collected

REPORT TITLE	DATE PUBLISHED	AUTHOR(S) or EDITOR(S)	ABSTRACT
Drainage Catchments in Major Cities			with properly coordinated instrumentation in networks representing a variety of climatic, topographic and land-use conditions. A properly designed information network is essential to secure maximum transferability of analysis results across a given region of the country and from one region to another. TM-10 summarizes the study which defines the size distribution and number of sewered drainage catchments in San Francisco, Washington DC, Milwaukee, Houston and Philadelphia. Tabulation of the sizes of all sewered drainage catchments, maps showing catchment boundaries and supporting discussion are included. The size distribution characteristics of the catchments varied considerably. The distribution of sewered drainage catchment size is unique for each city. There is no evidence at hand to indicate whether each city has distribution characteristics representative of its region or that each major city in the United States would have unique characteristics. The average catchment size varies from 65 acres in Houston to 560 acres in San Francisco. The upper quartile catchment size varies from 46 acres in Houston to 455 acres in San Francisco. The upper limit of drainage catchment size representing one half of the total city area served by sewers varies from 338 acres in Houston to 2,250 acres in Washington DC.
A Study of ASCE Urban Water Resources Research Council (UWRRC) - ASCE TM-8—	March-69	L.S. Tucker	The USGS Project on "Analysis of National Basic Information Needs in Urban Hydrology" is primarily aimed at improvement in design and management of urban storm drainage with a predominant emphasis on sewered catchments. Data are virtually non-existent that are suitable for the analysis of rainfall runoff-quality

REPORT TITLE	DATE PUBLISHED	AUTHOR(S) or EDITOR(S)	ABSTRACT
Availability of Rainfall-Runoff Data for Sewered Drainage Catchments			relationships, and that have been collected with properly coordinated instrumentation in networks representing a variety of climatic, topographic and land-use conditions. A properly designed information network is essential to secure maximum transferability of analysis results across a given region of the country and from one region to another.
			The purpose TM-8 is to make available to researchers information on available rainfall-runoff data for sewered drainage catchments. The authors hope that the data would meet an immediate need for initial testing of mathematical models developed previously for natural drainage catchments. Thirteen installations for the collection of rainfall and runoff data from completely sewered urban drainage catchments are presented. The study areas included Baltimore, Cincinnati, St. Louis, Chicago, Philadelphia, New York and Washington DC.
A Study of ASCE Urban Water Resources Research Council (UWRRC) - ASCE TM-9— *Raingage Networks in the Largest Cities*	March-69	L.S. Tucker	Considerable attention was given to several technical aspects of storm drainage. Major goals are: to arrive at general mathematical descriptions of the rainfall-runoff process combined with a basis for quantifying pollution loadings from storm sewerage systems applicable nationwide; to develop adequate technical knowledge for reliable planning of water quality and quantity exchanges between the several urban water service functions for multi-purpose development; and to facilitate and improve the planning, design and management of drainage works.
			The objectives of TM-9 are to inventory the extent of rain guage networks in the largest cities as guidelines for future data collection and to document the history of

REPORT TITLE	DATE PUBLISHED	AUTHOR(S) or EDITOR(S)	ABSTRACT
			data available for the information of researchers in urban hydrology. Only networks of automatic recording rain gauges were considered. The study focused on the 20 largest U.S. cities and their surrounding areas; the number of rain guages evaluated in any given city ranged from 1 to 162. The report comprehensively summarizes the nature of rainfall collection, analysis and interpretation methods in the varioius communities and the significance of the overall findings is discussed.
Report—*An Analysis of National Basic Information Needs in Urban Hydrology*	April-69	ASCE Staff	The *Analysis of National Basic Information Needs in Urban Hydrology* has focused on "data needs," "data devices" and "data networks." Primarily aimed at improvement in design of storm drainage, intensive study has been made of the data requirements for analyzing rainfall-runoff-quality relationships and of suitable data collection instrumentation, with consideration of the types of networks required for the collection of adequate data. Suitable data collected with properly coordinated instrumentation in networks representing a variety of climatic, topographic and land-use conditions, are virtually non-existent. There are very meager amounts of performance data with which existing or proposed storm drainage facilities can be checked or designed. Transfer of data findings between metropolitan regions is a central and primary objective. Urban water data acquisition and use would also be invaluable in facilitating the exchange of various quantities and qualities of water from one service to another, such as using stormwater for water supply. There are important implications in the use of data as part of the intelligence required for comprehensive management and operation

REPORT TITLE	DATE PUBLISHED	AUTHOR(S) or EDITOR(S)	ABSTRACT
			of urban water services.
ASCE Combined Sewer Separation Project TM-14—*Routing of Flows in Sanitary Sewerage Systems*	July-69	L.S. Tucker	The general concept on which the ASCE Combined Sewer Separation Project is based involves the discharge of comminuted or ground sewage from buildings and/or building complexes, via relatively small pressure tubing, into new pressure sanitary sewers. The new pressure sanitary sewers would discharge into existing interceptors that would convey the sanitary sewage to treatment works; stormwater would be conveyed in what were formerly combined sewers.
1969 Annual Convention of the Water Pollution Control Association of Pennsylvania—*Instrument Systems for Combined and Storm Sewer Research Data Acquisition*	August-69	L.S. Tucker	The continental United States is endowed with a rather fixed supply of fresh water. About 1,100 billion gallons discharge from its streams per day. Total withdrawals in 1954 amounted to some 300 billion gallons per day, and withdrawals that were later returned carrying some degree of pollution to water courses amounted to about 190 billion gallons per day. It is projected that in year 2000 total withdrawals will be about 4/5 the total stream flow and polluted returns will be about 2/3 the total stream flow.
			The effects and means of control of combined sewer overflows and stormwater discharges were inventoried by the APWA in 1967 in a project supported by the FWPCA. One of the several conclusions was that "there is a definite need for further research to determine the quantity and quality of overflows and the relative extent and detrimental effect of all sources of pollution." In a report to the Federal Council for Science and Technology the Committee on Pollution of the National Academy of Sciences stated that, "More research is needed to develop understanding of the whole stormwater pollution problem. This research should

REPORT TITLE	DATE PUBLISHED	AUTHOR(S) or EDITOR(S)	ABSTRACT
			cover the hydrology, the hydraulics, the treatment, the effects on receiving waters and related factors."
Combined Sewer Separation Using Pressure Sewers— *Feasibility and Development of a New Method for Separating Wastewater from Combined Sewer Systems*	October-69	ASCE	This report is concerned with the separation of community wastewaters, and runoff from rainfall and snowmelt in areas presently served by combined and intercepting sewers. Separation is accomplished by withdrawing the wastewater fraction of flows from existing plumbing systems and passing it through a sequence of added systems components as follows: (1) a storage, grinding and pumping unit within each building; (2) pressure tubing fished from the unit through each existing building sewer into the existing combined sewer; and (3) pressure piping inserted in that sewer and extending to the existing intercepting sewers that carry the wastewaters to treatment and disposal works. Runoff from rainfall and snowmelt, thus unencumbered by wastewaters, is removed from the community through the residual passageways of the one-time combined sewer system, which has thus become a combination of a new pressure conduit system within an old gravity conduit system.

The feasibility of this scheme of separation, the selection of available systems components and the development of required new systems components are described in this report on the basis of information drawn from 25 project reports and technical memoranda.

The feasibility of storing, grinding and pumping sewage from individual residences has been established; and standard comminuting and pumping equipment will be satisfactory for serving |

REPORT TITLE	DATE PUBLISHED	AUTHOR(S) or EDITOR(S)	ABSTRACT
			larger buildings. Acceptable types of pressure tubing are available that can be pushed and pulled through existing builidng drains and sewers. Pressure conduits can be suspended inside combined sewers that can be entered by workmen. There are combined sewer areas that can be separated most effectively by a version of the method investigated, but generally pressure systems will cost more than new gravity systems. New capabilities developed appear to be of potentially greater use for applications other than separation, such as new construction including utility corridors, and introduce viable alternatives for design of wastewater sewerage.
Report on the Third Conference on Urban Water Resources Research— *Urban Water Resources Management*	July-70	John R. Sheaffer, et al.	The first conference on urban water resources research (1965) probed many unknowns in the broad field of urban hydrology. Specifically, its focus was water and all of its relationships to urban man. A second conference (1968) addressed the interrelationships between man and his water resources. The conference reported herein was convened to explore broadly the many facets, dimensions, and problems of urban water resources management. This conference addressed research needs at interfaces with the total community, which remain only partially identified and which constitute generally unquantified "unknowns" that impede development of more rewarding and reliable approaches to urban water resource management organization and administration. Special attention was directed to the relationships of behavorial and instutional phenomena to physical phenomena. Approaches to and methods for stimulating and accelerating adoption of improved management

REPORT TITLE	DATE PUBLISHED	AUTHOR(S) or EDITOR(S)	ABSTRACT
			practices were also given particular attention. Considerable effort was devoted to a search for ways to make design and management more responsive to social and ecological needs and demands. There was further consensus that the problems presented in efforts to manage urban water resources were in large measure bounded by: (1) the consequences of professional efforts and activities with regard to the environment and the lives and life styles of human beings, (2) the need to obtain agreement upon the precise nature of objectives being sought, and (3) economic, institutional (including statutory), social and other constraints upon the adoption and implementation of worthwhile changes. These characteristics of the problem more or less define two major areas of data needs and research activity that must be pursued if greater benefits are to flow from our increasing technological refinement and its application. The engineer's role and the demands upon him are expanding. Today, an engineering design often must be considered unsatisfactory if it only maximizes or optimizes a client's needs or demands, or if it only satisfies some simple criterion such as a benefit/cost ratio. Today, it is growing steadily more necessary, and possibly completely necessary tomorrow, that the engineer and related scientists anticipate the consequences of designs or proposed technological changes upon others—the consequences for people and the environment upon which life depends. Successful adoption and implementation of designs and plans, moreover, are becoming increasingly dependent upon broader public concurrence in establishing and changing

REPORT TITLE	DATE PUBLISHED	AUTHOR(S) or EDITOR(S)	ABSTRACT
			goals and objectives.
A Study of ASCE Urban Water Resources Research Council (UWRRC) - ASCE TM-11—*Non-Metropolitan Dense Raingage Networks*	January-70	L.S. Tucker	Considerable attention was given to several technical aspects of storm drainage. Major goals are: to arrive at general mathematical descriptions of the rainfall-runoff process combined with a basis for quantifying pollution loadings from storm sewerage sewer systems applicable nationwide; to develop adequate technical knowledge for reliable planning of water quality and quantity exchanges between the several urban water service functions for multipurpose development; and to facilitate and improve the planning, design and management of drainage works. Considerable potential synergistic benefits could be obtained in comprehensive, multipurpose development of urban water, on a scale of involvement that in the aggregate could be much greater than for river basins. Whereas TM-9 documented the availability of rainfall data from networks in the largest cities, TM-11 indicates that an additional important source of rainfall data for research on metropolitan-scale storms is from dense rain guage networks located in non-metropolitan areas, which are summarized in TM-11. The authors evaluated 13 dense recording rain guage networks located in non-metropolitan areas of the U.S. Key information summarized in the study included: 1. Network name. 2. Network location 3. Organization Operating Network. 4. Equipment (number and kind of recording rain gauges). 5. Chart speed. 6. Approximate size of network. 7. Approximate average distance between gauges.

REPORT TITLE	DATE PUBLISHED	AUTHOR(S) or EDITOR(S)	ABSTRACT
			8. Period of record. 9. Intensity data 10. Form of reduced data. 11. Interpretation/significance.
A Study of ASCE Urban Water Resources Research Council (UWRRC) - ASCE TM-12— *Environmental and Technical Factors for Open Drainage Channels in Milwaukee*	February-70	Ted B. Prawdzik	Urban drainage systems represent large capital investments. The national replacement value of public storm and combined sewers has been estimated at $22 billion and the average annual public construction cost for urban storm drainage has been estimated at $2.5 billion for the next several years by the American Public Works Association. The AWPA also estimates that the national annual loss from urban drainage system flooding in recent years has been $1 billion or more.
A Study of ASCE Urban Water Resources Research Council (UWRRC) - ASCE TM-13— *Availability of Rainfall-Runoff Data for Partly Sewered Urban Drainage Catchments*	March-70	L.S. Tucker	The purpose of this technical memorandum is to identify available rainfall-runoff data for partially sewered urban drainage catchments to facilitate model development by researchers. Considerable detailed attention has been given to requirements for mathematical modeling of rainfall-runoff-quality in the first report by ASCE for OWRR. The very limited amount of available rainfall-runoff data for fully sewered catchments has been identified in an earlier technical memorandum. For the purpose of this technical memorandum a "partially" sewered drainage catchment is defined as one containing open channel water-courses together with developed secondary drainage systems consisting of street gutter drainage and/or roadside ditches and underground storm sewers.
Report— *Systems Analysis for Urban Water Management,*	September-70	Water Resources Engineers, Inc., Walnut Creek, CA	Expectations of private corporation management can be instructive by analogy. As industry managers seek to gain greater productivity through more integrated operations and the economies of scale,

REPORT TITLE	DATE PUBLISHED	AUTHOR(S) or EDITOR(S)	ABSTRACT
Prepared for the Office of Water Resources Research, U.S. Department of the Interior			reliability standards intensify as they try to insure the success of their enormous financial commitment; use of the computer as a routine tool in corporate decision-making is predicted; and under development is "an early-warning system for top management to see what is going into decisions rather than seeing results on financial reports at a point where it may be too late for correction." It must be clearly understood that the Steering Committee is <u>not</u> advocating the replacement of old techniques. Quite to the contrary, what is recommended is enhancement of management capability through the addition of new techniques which provide a greater insight with respect to the facilities and services for which managers are responsible. The objective is to maximize the value of what these managers already know. Outside specialists can assuredly <u>assist</u> managers in the type of development proposed, but the degree of direct involvement by upper management in the enterprise will determine its eventual success because the entire thrust of the exercise is management-oriented. The Steering Committee is of the opinion that implementation of the next phase of follow-on development proposed should materially assist the metropolitan authorities directly involved in the resolution of physical and financial expansion plans for their systems phased through the next several decades. The advantages of subsequent phases would accrue more to implementation and operation capabilities. The use of a comprehensive systems approach is advocated primarily as an overall means for

REPORT TITLE	DATE PUBLISHED	AUTHOR(S) or EDITOR(S)	ABSTRACT
			organizing the technical work and management decision-making processes, and as a framework for quantification techniques (whether the latter be information handling, simulation modeling, linear programming, cost-benefit analyses, or whatever). The adoption of the proposed overview approach certainly does not preclude or minimize the necessity or importance of the multitude of special-purpose techniques in common use—these will be unaffected except where the larger brush suggests modifications or accomodations.
Report— *Prospects for Metropolitan Water Management*	December-70	M.B. McPherson	ASCE is assisting in outlining and developing a coordinated national program of urban water resources research, including the provision of guidelines for long-range studies on urban water problems. The first phase of work was concluded with the publication of two reports. This study of metropolitan water management, including an assessment of related research needs, is a part of the second phase of work which is sponsored exclusively by the Office of Water Resources Research, US Department of the Interior. The objective was to conduct a study of metropolitan water management, taking into account: expectations for the environment of future cities; necessary or desirable new administrative arrangements for future public services management; and the expected impact and interaction of emerging technology, urbanization trends, social goals and related matters.
The June, 1971 Meeting of the Sub-Group on the Influence of Urbanization,	April-71	M.B. McPherson	As has been generally true around the world, U.S. institutions have not kept pace with the changing character of population distribution brought about by relentless urbanization, and urban hydrology has

REPORT TITLE	DATE PUBLISHED	AUTHOR(S) or EDITOR(S)	ABSTRACT
of the Working Group on the Influence of Man on the Hydrological Cycle; United Nations Educational, Scientific and Cultural Organization— *Preliminary Notes on the Hydrologic Effects of Urbanization in the United States*			accordingly suffered from inadequate attention. In sum, knowledge on the hydrologic effects of urbanization is fragmentary, and at a level inconsistently below its importance to society. More metropolitan-scale water-balance analyses should be undertaken as a means for improving overall water resources planning and management, and follow-on inventories should be made periodically to document change and to provide a better understanding of the effects of progressive urbanization. Research should be intensified on the fundamental rainfall-runoff-quality processes of urban catchments, for better planning, design and management of drainage systems, but particularly for the purpose of assessing the true generic role and character of pollution from discharges of separate underground storm drains and from combined sewer overflows. More adequate attention should be given to the translation of research findings into forms more usable in practical applications in order to realize greater public benefits therefrom.
A Study of ASCE Urban Water Resources Research Council (UWRRC) - ASCE TM-14— *Management Problems in Metropolitan Water Resource Operations*	September-71	M.B. McPherson	A recent ASCE Urban Water Resources Research Program report on metropolitan water management attempted to take into account: expectations for the environment of future cities; necessary or desirable new administrative arrangements for future public services management; and the expected impact and interaction of emerging technology, urbanization trends, social goals and related matters. Management tools have also been a Program concern.
A Study of	December-71	M.B.	This report is the last of three with an

REPORT TITLE	DATE PUBLISHED	AUTHOR(S) or EDITOR(S)	ABSTRACT
ASCE Urban Water Resources Research Council (UWRRC) - ASCE TM-15—*Feasibility of the Metropolitan Water Intelligence System Concept (Integrated Automatic Operational Control)*		McPherson	emphasis on urban water management. The first attempted to take into account: expectations for the environment of future cities; necessary or desirable new administrative arrangements for future public services management; and the expected impact and interaction of emerging technology, urbanization trends, social goals and related matters. The second dealt with operating problems and had as a major purpose the framing of a background for assessment of applicability and need of the intelligence system concept. Findings from these two reports are drawn upon for some of the conclusions reached in Section 3. The present report deals predominantly with the physical capability of existing technology.
Preliminary Notes on a Plan for an Urban Water Institute for Applied Development —*Urban Water Institute*	March-72	M.B. McPherson	Research on urban water resources around the world has received miniscule attention compared with non-urban counterparts. Root causes are the involvement of all levels of government in urban water matters and the balkanized character of local water agencies in almost every metropolis. Urban water management is an activity at the local level. Governments throughout the world exhibit the rural origins of their institutions, and the reality of the fact that many have a majority of their people in urban concentrations of population still has not been taken into accout. As a consequence, research conducted by national and territorial divisions (e.g., states), is framed from their perspectives and is available to urban officials as a zero-feedback input.
A Report for the AWWA Research Committee on Distribution Systems—	April-72	Edward E. Bolls, Jr., Dan A. Brock, Edwin B. Cobb, Holly A. Cornell, J.	This report summarizes water distribution research and applied development needs, as suggested by the AWWA research committee on distribution systems, through interaction with the UWRRC. Representative

REPORT TITLE	DATE PUBLISHED	AUTHOR(S) or EDITOR(S)	ABSTRACT
Water Distribution Research and Applied Development Needs		Ernest Flack, Frank Holden, F. Pierce Linaweaver, Jr., Robert C. McWhinnie, James C. Neill & Robert U. Olson	recommendations/observations included: 1. Using a number of systems, data sensitivity analysis should be made to better determine what kind and how much data should be collected for modeling purposes. 2. The time and spatial statistics of historical fire loads need evaluation on a national scale. 3. Hypothetical alternative designs of water services for segments of service districts representing residential occupancy should be prepared with the objectives of finding: the relationships between service pressures, back-up water facilities, building structural conditions fire-fighting capability as they affect fire protection and its cost; the incremental water system cost for meeting different peak rates and pressures; the relation between pressure requirements of appliances and pressure characteristics of plumbing compared with the cost of supplying those pressures, with a view towards reducing the combined costs of public services and private property; and the relation between lawn irrigation costs and reducing lawn irrigation load peaks. 4. The relationship between available pressure and required duration of lawn sprinkling should be investigated, together with an evaluation of pressure and duration desired by the public in a willingness to pay context. 5. There is wide variation in ratio of peak demand to average demand from one city to another and even between service districts within a

REPORT TITLE	DATE PUBLISHED	AUTHOR(S) or EDITOR(S)	ABSTRACT
			given city. 6. Better information on flow coefficients of pipes is needed, especially for smaller cities. 7. Water leakage represents lost revenue and the relationship between methods of construction and leakage should be examined in terms of newly developed materials and practices.
A Study of ASCE Urban Water Resources Research Council (UWRRC) - ASCE TM-16— *Metropolitan Industrial Water Use*	May-72	L.S. Tucker, Jaime Millan & Wilford W. Burt	Management of water resources on a comprehensive basis at the metropolitan level has been frustrated by the fact that the planning, implementation and operation of facilities and services are usually fragmented in both the central cities and in their metropolitan districts. Considerable attention has been accorded the modification of institutional arrangements for facilitating the management of metropolitan water resources, but these have been almost exclusively concerned with the direct jurisdictions of local government. The role of private interests was known to be substantial, but national water statistics have failed to identify metropolitan water resources as a separate entity, dwelling exclusively on river basin, state and national totals. This report may inspire a rectification of that oversight.
A Study of ASCE Urban Water Resources Research Council (UWRRC) - ASCE TM-17— *Hydrologic Effects of Urbanization in the United*	June-72	M.B. McPherson	The present challenge to the United States—and to every developed and developing nation—is to determine a more rational way of using resources so that economic growth and social progress can continue without jeopardizing the health, safety and well-being of people or endangering the Nation's security . . . . Only when confronted, in recent years, by gross pollution and threats of irreversible environmental damage have we begun to accept fully the fact that the wastes heedlessly generated by a growing,

REPORT TITLE	DATE PUBLISHED	AUTHOR(S) or EDITOR(S)	ABSTRACT
States			urbanized, high-production, high-consumption society exceed nature's capacity for self-renewal. . . . A new ethic has emerged which repudiates the mistakes of the past and demands the restoration and preservation of a safe, wholesome, aesthetically satisfying environment.[1] Environmental quality, in its broadest sense, has been defined as "a measure of all the physical, biological, social, economic, and cultural conditions and natural beauty which relate to the habitat of man and other creatures."[2] Priority environmental quality problem areas in the U.S. are water pollution, air pollution, long-term effects of human activity on climate, solid wastes disposal, noise, pesticides, radiation, and land-ues.[3] ───── [1] Department of State, U.S. National Report on the Human Environment, Department of State Publication 8588, GPO, Washington, DC, 1971, pp. 7, 17 & 18. [2] Stevenson, Albert H., "Human Ecology Considerations in Water Quality Management," in Is Water Quality Enhancement Feasible?, Specialty Conference Proceedings, ASCE, New York, 1970, pp. 66. [3] Council on Environmental Quality, Environmental Quality, GPO, Washington, DC, First Annual Report, Transmitted to the Congress August 1970.
ASCE Urban Water Resources Research Program TM-18—*Urban*	August-72	M.B. McPherson	This report deals with some of the principal aspects of surface runoff in urban areas in the U.S.

REPORT TITLE	DATE PUBLISHED	AUTHOR(S) or EDITOR(S)	ABSTRACT
Runoff			
Summary from Findings, Urban Water Resources Research Program, American Society of Civil Engineers— *Better Design of Stormwater Drainage Systems*	July-73	M.B. McPherson	What is now called the ASCE Council on Urban Water Resources Research held its first Engineering Foundation Research Conference in 1965 on *Urban Hydrology Research*. Among the principal 1965 findings was a serious need for field data on rainfall-runoff-quality for sewered catchments. Mainly in response to that expressed need, the ASCE Urban Water Resources Research Program was initiated late in 1967. A national program for acquisition of needed sewered catchment hydrological data was proposed to the US Geological Survey in 1969, together with a documentaiton of its justification, but progress in implementation has been painfully slow and future prospects unfortunately are not very bright, despite the best of intentions of all concerned. As of 1969: only a very few sewered catchments had been gauged; records seldom extended over more than a very few seasons; although rainfall-runoff had been measured, for still fewer of the catchments had rainfall-runoff-quality measurements been made; runoff measurements and the rarer runoff-quality measurements had been made at outfalls but no in-system; and of observations synchronization, and too crude signal resolution. Proportionally more rainfall-runoff data was available for partially sewered areas, but the amount was nevertheless pitifully small. Since then field research sponsored by EPA and other federal agencies has extended the database somewhat. However, not much new data has been acquired on completely sewered catchments using flumes or weirs rather than stage-guages for determining discharge.

REPORT TITLE	DATE PUBLISHED	AUTHOR(S) or EDITOR(S)	ABSTRACT
			In anticipation of the possible future availability of new data, considerations for modeling sewered urban catchment rainfall-runoff-quality processes were detailed and considerations for characterizing rainfall time and spatial distributions in future research were explored.
			The US Geological Survey is engaged in a program of data collection and studies to serve national needs in urban hydrology, but sewered catchment activities are limited. The Office of Water Resources Research has developed a broad program of projected urban water resources research, including urban hydrology analyses. Collection continues of information on the quality of storm sewer discharges and combined sewer overflows, supported by EPA.
DRAFT Part I, International Summary— *Hydrological Effects of Urbanization: Environmental Impact*	August-73	IHD/Unesco Sub-Group on the Effects of Urbanization on the Hydrological Environment	Among the numerous studies undertaken by the IHD was an investigation initiated in 1965 by the Working Group on the Influence of Man on the Hydrological Cycle, supported by the Food and Agricultural Organization (FAO), charged with agricultural and urbanization aspects of the subject. The Co-ordinating Council in 1970 decided to form a Sub-Group on the Effects of Urbanization on Hydrological Environment, supported by Unesco, to assist the Working Group and to investigate more intensively industrial and urbanization aspects.
DRAFT Part II, National Case Studies for Selected Countries— *Hydrological Effects of Urbanization:*	September-73	IHD/Unesco Sub-Group on the Effects of Urbanization on the Hydrological Environment	Chapter 1, Hydrologic Effects of Urbanization in the Federal Republic of Germany (by: Herbert Massing and Members of the State Institute for Hydrology and Water Protection Northrhine Westphalia Krefeld). Chapter 3, Hydrological Effects of Urbanization in the Netherlands (by: F.C. Zuidema,

REPORT TITLE	DATE PUBLISHED	AUTHOR(S) or EDITOR(S)	ABSTRACT
Environmental Impact			Ijsselmeerpolders Development Authority, Lelystad). Chapter 4, Hydrologic Effects of Urbanization in Sweden (by: Gunnar Lindh, Institutionen Fttenbyggnad Lund, Sweden). Chapter 6, Hydrologic Effects of Urbanization in the United States (by M.B. McPherson). Chapter 7, Hydrological Effects of Urbanization in the U.S.S.R. (by: V.V. Kuprianov, State Hydrological Institute Leningrad).
DRAFT Part III, Illustrative Special Topic Studies— *Hydrological Effects of Urbanization: Environmental Impact*	September-73	IHD/Unesco Sub-Group on the Effects of Urbanization on the Hydrological Environment	This draft of Part III has been prepared as a background document for the International Workshop on the Hydrological Effects of Urbanization, Warsaw, 8-10 November 1973. This deals with some of the principal aspects of surface runoff in urban areas of the U.S.A. Refer to the companion report, Chapter 6, Part II, for supplementary information related to the subject.
Journal of the Hydraulics Division— *Need for Metropolitan Water Balance Inventories*	October-73	M.B. McPherson	Paper of the Technical Council on Water Resources Planning and Management
A Study of ASCE Urban Water Resources Research Council (UWRRC) - ASCE TM-19— *Evaluation of Urban Flood Plains*	December-73	James E. Goddard	About 16 percent or 16,500 square miles of the nation's urban areas are in natural 100-year flood plains. This compares with 7 percent for the total national land area. Approximately 53 percent or 8,800 square miles of those urban flood plains have been developed. Better national and regional data inventories are needed to evaluate the flood problem. These are essential for preparation of wise water-related programs and also for guidance in industrial development, land use planning, transportation, environmental, and other programs.
Urban Land Institute, American	1974	Published Jointly by the Urban Land	As many miles of residential streets may be built in the next 30 years as have been built during the entire history of the United

REPORT TITLE	DATE PUBLISHED	AUTHOR(S) or EDITOR(S)	ABSTRACT
Society of Civil Engineers and National Association of Home Builders— *Residential Streets— Objectives, Principles and Design Considerations*		Institute, American Society of Civil Engineers and National Association of Home Builders—	States. Our streets have evolved from the often narrow dusty unpaved residential streets of 30 or 40 years ago to wide paved streets with curbs, gutters, sidewalks, street lighting, and landscaping. The process of converting from gravel or dirt streets to current practice has largely been the filtering down and adaptation of standards developed for highways having high volume heavy vehicle use. In an era when the national propensity was to see limitless national resources, the approach to setting standards often was "when in doubt, pick the highest standard."
			The purpose of this report is not to suggest that street design in the past has been "done all wrong" and should be completely redone, but rather to suggest objectives for our residential streets and principles to guide us toward optimum design standards. These objectives and principles are fundamentally common sense and should be readily acceptable. What is different about the objectives and principles is that they are specifically oriented toward residential streets, and they acknowledge that residential streets differ from commercial or industrial streets and from major arterials and interstate highways.
			It is presumptuous to suggest only one set of design standards could be uniformly applied throughout the country. At the same time, it is legitimate to suggest each community can measure its local practices against a set of objectives and principles which do have uniform applicability. Performance measured against these objectives and principles will identify the adequacy or inadequacy of current design and this will suggest modifications needed in local practice.

REPORT TITLE	DATE PUBLISHED	AUTHOR(S) or EDITOR(S)	ABSTRACT
Report of the Sub-group on the Effects of Urbanization on the Hydrological Environment, of the Coordinating Council of the International Hydrological Decade— *Hydrological Effects of Urbanization*	1974	M.B. McPherson	The objectives of this report are to describe the effects of urbanization together with its environmental impacts on the hydrological cycle and to recommend research needed by the managers of all types of water systems to minimize the environmental stresses. This report is primarily directed to researchers in hydrology, and a special summary is directed to water managers.
ASCE Urban Water Resources Research Program TM-20—*A Streamflow Model for Metropolitan Planning and Design*	January-74	Richard F. Lanyon & James P. Jackson, The Metropolitan Sanitary District of Greater Chicago	Mathematical models used for the simulation of urban rainfall-runoff or rainfall-runoff-quality can be divided into three distinct categories: Planning Models--Design Models--Operations Models.
Report to U.S. National Science Foundation— *International Workshop on the Hydrological Effects of Urbanization, Warsaw, 1973*	January-74	IHD/Unesco Sub-Group on the Effects of Urbanization on the Hydrological Environment	Among the numerous studies undertaken by the International Hydrological Decade was an investigation intitiated in 1965 by the Working Group on the Influence of Man on the Hydrological Cycle, supported by the Food and Agricultural Organization (FAO), charged with agricultural and urbanization aspects of the subject. The Coordinating Council in 1970 decided to form a Sub-Group on the Effects of Urbanization on the Hydrological Environment, supported by Unesco, to assist the Working Group and to investigate more intensively industrial and urbanization aspects.

REPORT TITLE	DATE PUBLISHED	AUTHOR(S) or EDITOR(S)	ABSTRACT
ASCE Urban Water Resources Research Program TM-21—*Innovation: A Case Study*	February-74	M.B. McPherson	"Society is always taken by surprise at any new example of common sense." Ralph Waldo Emerson. Innovation can be defined as: the act of introducing something new or novel, as in customs; a change effected by innovating; or a novelty added or substituted. The word derives from Latin and French terms for making anew or renewal. Thus, innovation can embrace the revisitation of an old idea, rephrased or reclothed for application in a new or different context. Related subjects include technology transfer, technology assessment, municipal delivery systems, municipal information systems, cost effectiveness evaluation, and public technology.
ASCE Urban Water Resources Research Program TM-22—*Computer Modeling Applications in Urban Water Planning*	March-74	Noel Hobbs & Jay D. Britton Denver Board of Water Commissioners	The purpose of this technical memorandum is to indicate the type of planning applications that have been made in a leading metropolitan water supply and distribution agency.
A Study of ASCE Urban Water Resources Research Council (UWRRC) - ASCE TM-23—*A Model for Evaluating Runoff-Quality in Metropolitan Master Planning*	April-74	L.A. Roesner, H.M. Nichandros, R.P. Shubinski, A.D. Feldman, J.W. Abbott & A.O. Friedland	Documented is a computer model that should see extensive use in total-jurisdiction preliminary sewerage planning, if for no reason other than the simple fact that it is presently the outstanding tool available for that purpose.
A Study of ASCE Urban	May-74	Water Resources	On February 5-9, 1973, a training course on "Management of Urban Storm Water,

REPORT TITLE	DATE PUBLISHED	AUTHOR(S) or EDITOR(S)	ABSTRACT
Water Resources Research Council (UWRRC) - ASCE TM-24— *Management of Urban Storm Runoff*		Engineers and The Hydrologic Engineering Center, Corps of Engineers	Quantity and Quality,"was sponsored by The Hydrologic Engineering Center, Corps of Engineers, U.S. Army, at its facilities in Davis, California. Lectures were involved with a quantity-quality mathematical model for preliminary planning called "STORM."
Proceedings of a Research Conference— *Urban Runoff Quantity & Quality*	August-74	M.B. McPherson, et al.	This conference was planned because the Urban Water Resources Research Council believed that insufficient progress was being made in developing adequate technology and data-gathering systems for urban drainage and runoff control. Public awareness of urban runoff problems is increasing, and governmental requirements for solutions are escalating. New technology which has been developed is only being sporadically applied. Therefore, the conference was directed at identifying technology gaps, research needs, and valuable new techniques which need only to be applied. The work of the Urban Water Resources Research Council, and of this conference in particular, covers topics inclusive of the interests of several of the ASCE technical divisions, and which also is of concern to disciplines other than civil engineering. Those invited to participate, although mainly civil engineers and administrators, included well-informed representatives of the social sciences and a few practicing environmentalists. The originally planned conference sessions successfully achieved most of the original objectives; but something more happened. As it proceeded, the conference developed a vitality of its own. Synergistic reactions occurred, and opinion started to coalesce as regards certain important points, not only

REPORT TITLE	DATE PUBLISHED	AUTHOR(S) or EDITOR(S)	ABSTRACT
			within the original agenda, but also outside of it. While some matters discussed are still unclear, and differences among individuals still prevail, there is a much larger degree of confidence that we are collectively on the right track, as results of experience and of research findings have combined to clear the focus of understanding. There still remains the task of expressing these new understandings in words clear enough and convincing enough to explain them to others. Hopefully, these proceedings will contribute to this broader understanding.
Engineering Foundation Conference (Paper)— *Urban Runoff, Quantity and Quality*	August-74	Remarks by M.B. McPherson	Local government agencies responsible for water resources are typified by a reactive, rather than an anticipatory, planning mode. Until there are more statutory or other kinds of incentives for anticipatory planning, innovative as well as traditional concepts will probably continue to be called upon on an ad hoc basis to meet each crisis as it arises. With all three levels of government on a reactive tack, each national shift causes perturbations through the State and then local governments, and the resultant policy instabilities have a pronounced inhibiting effect on the pursuit and acceptance of innovative programs. Too many researchers are inclined to believe that publications represent an exhaustive documentation of what is known, the state of the art. Instead, there is much more excellent innovative work that goes unreported and unnoticed. The fact that numerous "new" ideas turn out to be re-warmed or repeated concepts attests to this obtrusiveness. The scale and scope of investigation truly necessary for comprehensive analysis and resolution of problems and impediments in

REPORT TITLE	DATE PUBLISHED	AUTHOR(S) or EDITOR(S)	ABSTRACT
			the translation of research findings into practice, even when restricted to urban water applications, are enormous. Misinformation and lack of information can exacerbate conflicts, whereas rapprochements can be more readily reached when the public is in command of adequate, sound information.
A Study of ASCE Urban Water Resources Research Council (UWRRC) - ASCE TM-25—*ASCE Urban Water Resources Research Program, 1967-1974*	October-74	M.B. McPherson & G.F. Mangan, Jr.	Briefly summarized is the progress made through mid-1974, on research needs assessment, urban water management, translation of research findings into oractucem facilitation of urban runoff research, and collaboration and participation in research of local governments and other organizations. Also included are one-page abstracts for all 28 program reports and technical memoranda.
Urban Land Institute, American Society of Civil Engineers and National Association of Home Builders— *Residential Stormwater Management— Objectives, Principles and Design Considerations*	1975	Published Jointly by the Urban Land Institute, American Society of Civil Engineers and National Association of Home Builders	In the next 25-years, more money may be spent for stormwater management than has been spent for drainage during the entire history of the Nation. This will amount to billions of dollars, not including funds that will be expended to maintain water quality, which also is an increasing national concern. Much, if not most, of this investment will be private funds expended during the development of land for urban uses. Ultimately, however, all costs (capital, debt and maintenance) are borne by all citizens. Public investments should be made wisely in furtherance of the quality of life so high valued. It is the opinion of ASCE, NAHB, and ULI that the application of the Objectives, Principles, and Design Considerations contained in this report, with due regard for unique and particular circumstances and conditions found in various areas of the country, will

REPORT TITLE	DATE PUBLISHED	AUTHOR(S) or EDITOR(S)	ABSTRACT
			be a significant step in the right direction.
			For the concepts in the report to achieve wide application, there will be a need to induce institutional changes even beyond the design professions and the regulatory institutions of governments. Changes also are necessary in the financial institutions which fund development based on their approval of a project's design, in the insuring institutions and their perceptions of the insurability of this approach, and in the legal professions in relationship to the pre-existing body of land use law regarding rights, responsibilities and liabilities. The implication of the above statement, that change may be precluded by institutional constraints, is real. The importance of encouraging application of worthwhile new approaches should provide the impetus to achieve necessary changes.
			It is hoped that this report will motivate creative rethinking and updating of drainage design practices. Anyone disagreeing with any part of this report is encouraged to advise the publishers of that disagreement, including their detailed reasons and alternative recommendations. Such information will help guidie future revisions and enhance the document's value to our Nation.
A Study of ASCE Urban Water Resources Research Council (UWRRC) (First Draft for March 1975 Panel Review at USGS,	February-75	M.B. McPherson	The San Francisco Bay Region Environment and Resources Planning Study has been conducted by the Geological Survey of the U.S. Department of the Interior under a joint sponsorship with the U.S. Department of Housing and Urban Development. The primary goal of the Study was to provide the kinds of earth science data that would be of material assistance in regional urban planning and development, particularly information

REPORT TITLE	DATE PUBLISHED	AUTHOR(S) or EDITOR(S)	ABSTRACT
Menlo Park, California— *Earth Science Information Needs in Local Government Water Control Master Planning*			regarding rock and soil conditions, slope stability, availability and quality of water, susceptibility to earthquake damage, waste disposal and storage, and locations of construction materials and other resources.
ASCE Urban Water Resources Research Program TM-26—*Effects of Urbanization on Water Quality*	March-75	Robert P. Shubinski and Steven N. Nelson	The basic content of this technical memorandum is based heavily on a paper presented at the ASCE Urban Transportation Division's Specialty Conference on "Environmental Impact" held May 21-23, 1973, in Chicago. While the Proceedings of the conference have been published recently by ASCE (*Environmental Impact*, ASCE, New York, NY, 393 pp., 1975), the original paper is not likely to be sought therein by a large enough number of urban planners. In order to reach more urban planners, Messrs. Shubinski and Nelson have rewritten and enlarged the scope of the conference paper as a voluntary contribution to the ASCE Program. Erosion evaluation criteria being used in the development of the Fairfax County, Virginia, Master plan supplement those outlined in this technical memorandum.
A Staff Report for the NSF-RANN Project G 42282 "Petroleum Industry in the Delaware Estuary"— *Projection of Petroleum Content of Urban Runoff*	April-75	William Whipple, Jr., Joseph V. Hunter, Shaw L. Yu	The first ten months of data gathering and analysis indicate an appreciable content of petroleum in storm sewers. If the concentration found elsewhere proves to be about at this level, the petroleum in urban runoff into the Delaware Estuary will be a source of considerably more petroleum than the effluents of the seven refineries of the estuary, after secondary treatment.

REPORT TITLE	DATE PUBLISHED	AUTHOR(S) or EDITOR(S)	ABSTRACT
A Study of ASCE Urban Water Resources Research Council (UWRRC) - ASCE TM-IHP-1—*Urban Mathematical Modeling and Catchment Research in the U.S.A*	June-75	M.B. McPherson	This report is a U.S. contribution to Unesco state-of-the-art reports on urban hydrology as part of the International Hydrological Programme.
Report— *Regional Earth Science Information in Local Water Management (Earth Science Information Needs in Local Government Water Control Master Planning)*	July-75	M.B. McPherson	The major USGS-ASCE project objective was to make an assessment of the usefulness of information products of the type generated by the SFBRS, in their existing form or as modified or extended, for master planning of water resource capital improvements at the local level of government. The relationship of data needs for local government urban water management to the more than 80 products of the SFBRS is evaluated in the last part of the report, and is mostly the contribution of a volunteer ad hoc Review Panel convened for that purpose. Background for the findings on national transferability occupies the main body of the report, in which are discussed: master planning, urban water management and planning interactions; selected components of urban water management; selected techniques in urban water management; and three selected case studies. A large proportion of the SFBRS type of products has been adjudged usable in other metropolitan areas, in their present form or in a more detailed form, for local government master planning. The report closes with an entreaty for more extensive

REPORT TITLE	DATE PUBLISHED	AUTHOR(S) or EDITOR(S)	ABSTRACT
			and widespread use of such information.
Journal of the Hydraulics Division— *ASCE Urban Water Resources Research Program*	July-75	M.B. McPherson & George F. Mangan, Jr.	ASCE UWRRC have focused the attention of public agencies at all levels of government on critical research needs. Recognizing the casm that had to be bridged, the Council formed the ASCE Program in 1967 to serve as its temporary full-time operating arm. Twenty-four technical memoranda of value to all water resource practitioners in the urban field have been prepared with the help of outside contributors.
Regional Earth Science Information In Local Water Management— *Study of ASCE Urban Water Resources Research Council*	July-75	M.B. McPherson	Prepared on behalf of the ASCE URBAN Water Resources Research Council, the following project report on Earth Science Information Needs in Local Government Water Control Master Planning was prepared for the San Francisco Bay Region Environment and Resources Planning Study (SFBRS), U.S. Geological Survey Experimental Program designed to facilitate the use of earth-science and related information in metropolitan planning and decision-making.
A Study of ASCE Urban Water Resources Research Council (UWRRC) - ASCE TM-IHP-1—*Urban Mathematical Modeling and Catchment Research in the U.S.A.*	November-75	M.B. McPherson	A preliminary version of this report, dated June, 1975 was distributed to a number of program co-operators. Most of their suggestions have been taken into account in this report. The report is a U.S. contribution to Unesco state-of-the art reports on urban hydrology as part of the International Hydrological Program. ASCE took early supportive action by applying for an NSF grant to assist in two of the ten recommended projects: R1. <u>Catchment Studies Report</u>. Perpare a "state-of-the-art" report on research executed in urban catchment areas, which would include instrumentation, data

REPORT TITLE	DATE PUBLISHED	AUTHOR(S) or EDITOR(S)	ABSTRACT
			acquired, analysis performed and applications.
			R3. <u>Mathematical Models Report</u>. Prepare a "state-of-the-art" report on mathematical models applied to urban catchment areas and dealing with, for instance, rainfall-runoff relationships and water balances, both with respect to water quantity and quality.
			In keeping with the findings of the Subgroup and the Warsaw Workshop, <u>modeling and catchment research for urban drainage systems is emphasized</u>. This is the subject singled out as having the largest gaps in knowledge in urban hydrology. Water quality aspects are accorded considerable attention because of the strong interest in environmental protection in the U.S.
ASCE Urban Water Resources Research Program TM-27— *Commercial Water Use (Water Use in Selected Commercial and Institutional Establishments in the Baltimore Metropolitan Area)*	December-75	Jerome B. Wolff, F.P. Linaweaver, Jr. and John C. Geyer, The Johns Hopkins University, Baltimore, MD	Priorities in Distribution Research and Applied Development Needs Committee Report. A committee report presented at the Annual Conference on June 14. Since the publication of the committee report, Water-Distribution Research and Applied Development Needs, in the June 1974 JAWWA, two subcommittees have prepared statement highlighting urgently needed work within two previously defined priority areas. Those statements are included in this article in addition to comments offered by committee members.
OKLA C-5228 Agreement No. 14-31-0001-	September 1973 - March 1976	James E. Garton & Charles Rice	In temperate climates, many lakes stratify during the summer. A typical stratified lake will have a warm oxygen-rich

REPORT TITLE	DATE PUBLISHED	AUTHOR(S) or EDITOR(S)	ABSTRACT
4215, submitted to The Oklahoma Water Resources Research Institute—*Improving the Quality of Water Releases from Reservoirs by Means of a Large Diameter Pump*			epilimnion, a thermocline, and a colder oxygen depleted hypolimnion. High levels of iron, manganese, hydrogen sulfide, and ammonia and dissolved hydrocarbons may occur in the hypolimnion. Many efforts have been made to destratify lakes, primarily by air bubbling. These methods requre large energy inputs. A low-energy destratifier using 42-inch and 72-inch propellers to pump the water downward from the surface have been used successfully to destratify a 100 acre lake (35 feet deep) for 3 years. This project was an attempt to apply the same kind of device to Lake Arbuckle, a 2400 acre, 90-foot deep lake in south central Oklahoma. A 16.5 foor aricraft propeller was used to pump approximately 200,000 gallons per minute downward. Although the lake stability index was decreased by half, a corresponding reduction in the oxygen distribution index did not occur until the fall overturn. Thus, a lake can be weakly stratified thermally and strongly stratified chemically. The fall turnover occurred about a month earlier and more completely than normal during our two years of operation. The oxygen content in the outlet waters near the pump was increased 1 to 2 mg/L during the critical summer months.
ASCE Urban Water Resources Research Program TM-28—*Household Water Use*	January-76	M.B. McPherson	Primarily addressed to researchers, the objectives of this report on household water use are definitions of: the status of information; changes in public policies and goals; and consequent needs for new information and research. Discussed are some of the more important quality, quantity and economic considerations. Demand variations are given predominant attention, for individual and multiple households. An appendix contains a listing of a unique set of water-use data for nine individual households covering a two week

REPORT TITLE	DATE PUBLISHED	AUTHOR(S) or EDITOR(S)	ABSTRACT
			period at a one-minute recording interval.
ASCE Urban Water Resources Research Program TM-29—*Computerized City-Wide Control of Urban Stormwater*	February-76	Neil S. Grigg, John W. Labadie, George R. Trimble, Jr., & David A. Wismer	A comprehensive research program on automation of urban stormwater facilities has been undertaken at Colorado State University. With the support of the Office of Water Research and Technology, automation implications for individual catchments were investigated and managerial aspects of the findings were reported separately from technical project completion reports to enhance their transferability. With the support of the Research Applied to National Needs Program of the National Science Foundation, automation implications for total-jurisdictiion systems were investigated, the scope of the report that follows.
A Report for Division of Advanced Environmental Research and Technology Research Applied to National Needs Program National Science Foundation Washington, DC 20550—*Urban Hydrology Research Needs*	April-76	M.B. McPherson	Hydrology may be defined as the science that is concerned with the waters of the earth, their occurrence, circulation and distribution, chemical and physical properties, and their reaction with the environment, including their relationship to living things. Thus, hydrology embraces the full history of water on the earth. Because the complex interactions of human activity in concentrated settlements with air, water and land must be taken into account, urban hydrology is a distinctive branch of the broad field of hydrology. As opposed to conventional hydrology, because urban development everywhere has been in continuous states of expansion and flux, urban hydrology contends with the dimension of dynamic change. Also, urban water resources management utilizes the social and biological sciences as well as the physical sciences. Reflecting the lag in the recognition of the fact that America became an urban nation over half a century ago, the term urban hydrology gained

REPORT TITLE	DATE PUBLISHED	AUTHOR(S) or EDITOR(S)	ABSTRACT
			currency less than two decades ago. Since then, the term has been tacitly expanded to include all urban water resource matters in, or interfacing with, the hydrologic cycle, including water quality considerations. In 1972, reference was made to "the continuing ambivalence of the Federal Government on the question of what its real role is in our nation's metropolitan areas," a status that has changed little since, witness the fiscal dilemma of New York City and other urban centers. All levels of government are commonly involved to some degree in urban water resources research, and the fractionalized character of local government at the metropolitan level adds to the diffusion of attention. In sum, "even the most urbanized countries still exhibit the rural origins of their institutions." Few local agencies can support hydrologic research that will yield results transferrable to other metropolitan areas, or even from one jurisdiction to another in the same metropolis. Hence, urban water resources research around the world commonly has suffered from inadequate attention and support, and from discontinuous and erratic efforts. More than a decade ago it was possible to say that there was a need for research to establish the nature of the effects of urbanization on basic hydrological processes. Today, the broad nature of these effects is only beginning to be understood.
A Study of ASCE Urban Water Resources Research Council (UWRRC) - ASCE TM-	May-76	A.P. Aitken	This report is the second in a special series of ASCE Program Technical Memoranda for the International Hydrological Programme.

REPORT TITLE	DATE PUBLISHED	AUTHOR(S) or EDITOR(S)	ABSTRACT
IHP-2—*Urban Hydrological Modeling and Catchment Research in Australia*			
A Study of ASCE Urban Water Resources Research Council (UWRRC) - ASCE TM-IHP-3—*Urban Hydrological Modeling and Catchment in Research in Canada*	June-76	J. Marsalek (Canada Centre for Inland Waters Burlington, Ontario, Canada)	This endeavor originated from activities and aspirations of the Unesco Subgroup on the Hydrological Effects of Urbanization of the International Hydrological Decade which concluded in 1974. Members of the the Subgroup represented the Federal Republic of Germany, France, Japan, Netherlands, Sweden, U.K., U.S.S.R. and U.S.A.
ASCE Urban Water Resources Research Program TM-30—*Urban Flood Warning and Watershed Management Advances in Metropolitan Melbourne*	June-76	C.T. Earl, J.W. Porter, J.A. Lanaway, A.S. Alexander, R.W.M. King, G.O. Cosgriff--Melbourne & Metropolitan Board of Works and Don G. Thompson--Dandenong Valley Authority	In November, 1975 M.B. McPherson was privileged to present a paper in Canberra at the third National Symposium on Hydrology sponsored by the Australian Academy of Science. Professor E.M. Laurenson of Monash University arranged for a visit to his university, the University of Melbourne, the Melbourne & Metropolitan Board of Works and the Dandenong Valley Authority, all in metropolitan Melbourne, at the close of the symposium in Canberra. Intrigued by advances made in institutional arrangements and urban flood warning systems in Melbourne, the authors of this report were asked to contribute suitable material for an ASCE Program Technical Memorandum.
A Study of ASCE Urban Water Resources Research	July-76	Edited by M.B. McPherson	Proceedings of a Special Session, Spring Annual Meeting, American Geophysical Union, Washington DC April 14, 1976. Over the past few years there has been an unprecedented surge of activity in urban

REPORT TITLE	DATE PUBLISHED	AUTHOR(S) or EDITOR(S)	ABSTRACT
Council (UWRRC) - ASCE TM-31—*Utility of Urban Runoff Modeling*			hydrological model development. While there has always been a lag between development and application, many specialists in urban water resources matters have been concerned about what has seemed to be a snail-pace adoption of the newer tools. There is widespread agreement that what is hindering more extensive and effective use of more sophisticated techniques in local government projects is a clear definition of why their use would be more cost-effective than simpler, traditional procedures.
A Study of ASCE Urban Water Resources Research Council (UWRRC) - ASCE TM-IHP-4—*Urban Hydrological Modeling and Catchment Research in the United Kingdom*	July-76	M.J. Lowing	This report is the fourth in a special series of ASCE Program Technical Memoranda for the International Hydrological Programme. The prototype report was for the U.S.A. A number of contributed national reports, of which the United Kingdom report herein is the third to be received, will be issued subsequently through 1976.
A Study of ASCE Urban Water Resources Research Council (UWRRC) TM-IHP-6—*Urban Hydrological Modeling and Catchment Research in the United Kingdom*	July-76	M.J. Lowing	This endeavor originated from activities and aspirations of the Unesco Subgroup on the Hydrological Effects of Urbanization on the International Hydrological Decade which concluded in 1974. This report summarizes urban hydrologic modeling and catchment research in the United Kingdom.

REPORT TITLE	DATE PUBLISHED	AUTHOR(S) or EDITOR(S)	ABSTRACT
ASCE Urban Water Resources Research Program TM-31—*Utility of Urban Runoff Modeling*	July-76	M.B. McPherson	Proceedings of a special session, Spring Annual Meeting, American Geophysical Union, Washington DC, 14 April 1976.
Supplement to the Technical Completion Report Project C-5228—*Demonstration of Water Quality Enhancement Through the Use of the Garton Pump*	August-76	James E. Garton & Howard R. Jarrell	A field demonstration of state-of-the-art technology was held at Lake Okatibbee near Meridian, MS. This demonstration included a review of reservoir stratification dynamics and of the development and previous research involving the Garton Pump. Upon conclusion of the "chalk talk," participants adjourned to the lake intake structure and stilling basin to observe the pump operations at first hand.
A Study of ASCE Urban Water Resources Research Council (UWRRC) TM-IHP-5—*Methods for Calculating Maximum Flood Discharges for Natural Watercourses and Urban Areas in the U.S.S.R.*	August-76	V.V. Kuprianov	This endeavor originated from activities and aspirations of the Unesco Subgroup on the Hydrological Effects of Urbanization on the International Hydrological Decade which concluded in 1974. This report summarizes methods for calculating maximum flood discharges for natural watercourses and urban areas in the USSR.
A Study of ASCE Urban Water Resources Research	September-76	H. Massing	This endeavor originated from activities and aspirations of the Unesco Subgroup on the Hydrological Effects of Urbanization on the International Hydrological Decade which concluded in 1974. This report

REPORT TITLE	DATE PUBLISHED	AUTHOR(S) or EDITOR(S)	ABSTRACT
Council (UWRRC) TM-IHP-6— *Urban Hydrology Studies and Mathematical Modeling in the Federal Republic of Germany*			summarizes urban hydrologic studies and mathematical modeling in the Federal Republic of Germany.
A Study of ASCE Urban Water Resources Research Council (UWRRC) TM-IHP-7— *Urban Hydrological Modeling and Catchment Research in Sweden*	October-76	Gunnar Lindh	This endeavor originated from activities and aspirations of the Unesco Subgroup on the Hydrological Effects of Urbanization on the International Hydrological Decade which concluded in 1974. This report summarizes urban hydrologic modeling in catchment research in Sweden.
A Conference Report—*Guide for Collection, Analysis, and Use of Urban Stormwater Data*	November-December-76	William M. Alley	The objective of this report is to present guidelines for designing a data-collection program to generate, within the context of available resources, data of sufficient breadth, detail and representativeness to define urban-runoff problems, to calibrate and verify urban-runoff models, and to monitor the effectiveness of "solutions" to urban-runoff problems. Rising project complexity, increasing potentials for investments required, and continually changing "rules of the game" such as the 1972 Amendments to the Federal Water Pollution Control Act, the flood insurance acts, more extensive land-use planning, the environmental movement, etc., are all resulting in an ever-increasing

REPORT TITLE	DATE PUBLISHED	AUTHOR(S) or EDITOR(S)	ABSTRACT
			need for more field data. Local governments are essentially on their own for the acquisition of local urban-runoff field data. Documentation of the methodology for acquiring and analyzing urban-runoff data is scattered throughout the literature and guidelines for developing a complete "data collection package" are not presently available. This publication attempts to fill that gap. The emphasis is not on instrumentation theory, laboratory analytical techniques or model theory, but rather on guidelines that can be used by local governments for an approach to data collection that is based on the purposes for which the data are to be used. Five separate, but interdependent topics are discussed: data utilization, data analysis, network planning and design, instrumentation, and data collection. Summaries of these chapters are presented in Table 1.
Article Reprinted from EOS, Volume 57, No. 11, November 1976—*Urban Water Resources*	November-76	M.B. McPherson	Urbanization growth.
Engineering Foundation Conference— Eason, MD "NETWORK DESIGN" (emphasis on temporal factors)— *Instrumentatio n and Analysis of Urban*	November-76	M.B. McPherson	The purpose of this presentation is to supplement or compliment the coverage in the workshop report draft issued November 28 to participants, with regard to temporal aspects of data network design. This workshop has its origins in the International Hydrological Programme (IHP). Project 7 of the IHP, *Effects of Urbanization on the Hydrological Regime and on the Quality of Water*, contains five sub-projects, one of which is *Research on*

REPORT TITLE	DATE PUBLISHED	AUTHOR(S) or EDITOR(S)	ABSTRACT
Stormwater Data, Quantity and Quality			*Urban Hydrology* (sub-project 7.1, for which the speaker is the rapporteur for Unesco). Included as the initial effort of sub-project 7.1 is a series of national case studies, which are being assembled as ASCE Program Technical Memoranda in a special IHP series (processed to date are reports for the USA, Australia, Canada, United Kingdom, USSR, Federal Republic of Germany, Sweden and France). In addition, sub-project 7.1 calls for the preparation of information manuals on urban water-data collection, analysis and use, with the first such manual emphasizing catchments. Having failed to obtain financial support for an international symposium to develop a catchment manual, and recognizing the urgent need for such a manual in the US, the ASCE UWRRC turned instead to the development of the present workshop. An international accent has been maintained by the inclusion of national reports in the workshop program for Canada and Sweden.
ASCE Urban Water Resources Research Program TM-IHP-8—*Urban Hydrological Modeling and Catchment Research in France*	November-76	M. Desbordes & D. Normand	This report has been prepared on behalf of the French National Committee for the International Hydrological Programme as a contribution to IHP Project 7, "Effects of Urbanization on the Hydrological Regime and on Quality of Water." Reviewed in this section are the present status of research on urban hydrology in France and plans for future research in that field.
ASCE Urban Water Resources Research Program TM-IHP-9—*Urban Hydrological Modeling and*	December-76	Nils Roar Selthun	This report is a Norwegian contribution to the UNESCO state-of-the-art report series on urban hydrology (IHP Subprojects 7.1 and 7.2) The report has been prepared under a project on urban catchment research, Project PRA 4.2, financed by the Norwegian Ministry of Environmental Affairs and administered by the

REPORT TITLE	DATE PUBLISHED	AUTHOR(S) or EDITOR(S)	ABSTRACT
Catchment Research in Norway			Hydrological Division of the Norwegian Water Resources and Electricity Board.
A Study of ASCE Urban Water Resources Research Council (UWRRC) - ASCE TM-IHP-10— *Urban Hydrological Modeling and Catchment Research in the Netherlands*	January-77	F.C. Zuidema	In spite of a continuous growth in the population of the Netherlands (13 million in 1970 with 15.6 million expected in 2000) and an increasing population density (384 inhabitants per square km in 1970, 403 in 1975, 434 in 1985, and 463 in 2000), development in urban water resources research has been rather tardy. However, there is an enormous diversity among urban hydrological problems, which are solved adequately. For example, while only a few urban catchment studies are going on, mathematical models are used for different goals and at different levels: models for rainfall-runoff, water quality, comprehensive urban water and water resources management.
Paper—*The Design Storm Concept*	February-77	M.B. McPherson	Historically, urban areas have been drained by underground systems of sewers that were intentionally designed to remove stormwater as rapidly as possible from occupied areas. Discharges from conventional storm drainage sewer facilities and floodplain intrusion by structures both tend to aggravate flooding, and thereby jointly tend to raise the potential for stream flooding damages. The advantages of local detention storage in lieu of the traditional rapid removal of storm flows has long been recognized, and there is evidence that the usage of such storage is on the rise. However, local detention storage has been only occasionally employed as part of overall flood mitigation, as in Denver, Coloardo and Fairfax County, Virginia.

Detention storage is recognized as one of the principal means for abatement of pollution from stormwater discharges and |

REPORT TITLE	DATE PUBLISHED	AUTHOR(S) or EDITOR(S)	ABSTRACT
			combined sewage overflows. There are opportunities in new land development to incorporate detention storage at the ground surface. However, for existing drainage systems, there may be few opportunities to add detention storage except underground, for both combined sewer and separate storm sewer systems. Because abatement of pollution from the dispersed sources served by drainage systems began with a focus on combined systems, our knowledge on storage requirements for them is greater. Recognized very early was the need for some for of automatic control because of system complexity and the need to manipulate flows in order to insure their containment as a means for reducing overflows.
A Study of ASCE Urban Water Resources Research Council (UWRRC) - ASCE TM-IHP-11— *Urban Runoff Research in Poland*	February-77	Pawel Blaszczyk	Earlier research in Warsaw on the influence of combined sewerage schemes on the Vistula River, as well as a survey of the area conducted as part of a general project for development of sewerage and drainage in Warsaw, clearly showed the need for investigation of the relations between rainfall and the runoff from drainage catchments in order to meet the requirements for planning and designing sewerage and drainage schemes.
A Study of ASCE Urban Water Resources Research Council (UWRRC) - ASCE TM-IHP-12— *Urban Hydrological Modeling and Catchment*	May-77	S. Ramaseshan & P.B.S. Sarma	This report summarizes the state of the hydrology practice in India. Because of the low priority for urban hydrology in India, and scanty information dealing with mathematical models and urban hydrologic research, the scope of this report is extremely limited.

REPORT TITLE	DATE PUBLISHED	AUTHOR(S) or EDITOR(S)	ABSTRACT
Research in India			
A Study of ASCE Urban Water Resources Research Council— *Urban Runoff Control Planning*	June-77	M.B. McPherson	The growth impact of projected areal enlargement of urban areas on present and planned urban water resource facilities is almost too stunning a reality to be comprehended fully at this time. Even if expected urban growth is checked by a renaissance of our central cities, the required reconstruction will still be monumental.
A Study of ASCE Urban Water Resources Research Council (UWRRC)— *Urban Runoff Control Planning*	June-77	M.B. McPherson	Section 208 of Public Law 92-500 (Federal Water Pollution Control Act of 1972) encourages areawide planning for water pollution abatement management, including urban runoff considerations where applicable. This report has been prepared to assist agencies and their agents that are participants in the preparation of areawide plans, from the standpoint of major urban runoff technical issues in long-range planning.
A Study of ASCE Urban Water Resources Research Council (UWRRC) - ASCE TM-IHP-13— *Urban Hydrological Modeling and Catchment Research: International Summary*	November-77	M.B. McPherson & F.C. Zuidema	Twelve national reports for the International Hydrological Programme (IHP) have been compiled on the state-of-the-art in urban catchment research and hydrological modeling, with particular attention given to underground conduit systems. Summarized in this report are their principal commonalities, together with particularly noteworthy observations or advances reported for individual countries.
ASCE Urban Water Resources Research Program TM-	December-77	William H. Espey, Jr.; Duke G. Altman & Charles B.	Unit hydrographs have been developed from rainfall-runoff field data for completely sewered drainage catchments in Louisville, KY and Atlanta, GA. However, there is considerable doubt over the

REPORT TITLE	DATE PUBLISHED	AUTHOR(S) or EDITOR(S)	ABSTRACT
32— *Nomographs for 10-Minute Unit Hydrographs for Small Urban Watersheds*		Graves, Jr.	transferability or universality of the findings from a single jurisdiction. Moreover, most of the unit hydrographs developed across the nation have been for partially sewered catchments where the streamflows measured had included a significant contribution from non-sewered sectors. An example is the testing of unit hydrographs using urban streamflow data from large catchments nearby by the Georgia Institute of Technology.
ASCE Paper— *Urban Runoff Control, Quantity and Quality*	March-78	M.B. McPherson	Presented at American Public Works Association Urban Drainage Workshop, Omaha, Nebraska March 14, 1978. This paper summarizes the main points that have been covered in presentations to areawide planning agencies regarding urban runoff management.
Proceedings of the Research Conference— *Water Problems of Urbanizing Areas*	July-78	William Whipple, Jr., et al.	This conference grew out of continuing concern by various members of the Universities Council on Water Resources and of the Urban Water Resources Research Council, ASCE, during 1977 that the water problems of urbanizing areas were not being given sufficient attention. Originally, the conference was intended to give principal attention to water quality and environmental problems, covering other water problems primarily as interfaces. However, the sponsoring agencies urged broadening of the scope; and, accordingly, water supply and flood control problems also were given attention. This change was desirable, but it considerably increased the scope of coverage of the conference. Please note the distinction between "water problems of urbanizing areas" and "urban water problems." The latter has been the subject of considerable research and is a problem of substantial economic importance. The subject of this conference is intended to be more focused,

REPORT TITLE	DATE PUBLISHED	AUTHOR(S) or EDITOR(S)	ABSTRACT
			concentrating on the unique problems of those areas undergoing the process of urbanization. Our theme this week is "Water Problems of Urbanizing Areas." This concept is significantly different from the urban water programs and conferences of the past few years, in that we are emphasizing metropolitan regions rather than central cities, and we are considering a dynamic process rather than a condition. In treating of environmental pollution in urbanizing areas, we will be thinking of the planning processes which may limit and control development, to the aim that areas not yet fully developed may be able to preserve their environmental amenitieis. In the past, some of us tried to recognize the broader planning aspect by stressing water problems of metropolitan rather than urban areas; but this concept was too fine a distinction and never had much appeal except to theoreticians. The term "urbanizing areas" brings out the dynamics of the situation, where the problems are most acute and the necessities for better management most apparent.
A Study of ASCE Urban Water Resources Research Council (UWRRC) - ASCE TM-33—*Research on the Design Storm Concept*	September-78	Jiri Marsalek	This Technical Memorandum is Addendum 4 of a 1977 ASCE Program report on "Urban Runoff Control Planning." Addendum 1, "Metropolitan Inventories," and Addendum 2, "The Design Storm Concept," were appended to the latter report. Addendum 3 was the first of several additional, individual Addenda to be released over the period 1977-1979.
A Study of ASCE Urban Water Resources	October-78	Jess Abbott	Six models, plus two variants of one and a variant of another, were tested in this study, with the objective of making a preliminary evaluation of their relative capabilities,

REPORT TITLE	DATE PUBLISHED	AUTHOR(S) or EDITOR(S)	ABSTRACT
Research Council (UWRRC) TM-34— *Testing of Several Runoff Models on an Urban Watershed*			accuracies and ease of application. A detailed comparison of the many capabilities and features of these models was beyond the scope of the study. For four of the models, plus two variants of one of them, the primary performance criterion was the degree to which simulated values matched observed daily and monthly runoff volumes for the 5.5-square mile Castro Valley Watershed near Oakland, California.
A Study of ASCE Urban Water Resources Research Council (UWRRC) - ASCE TM-35—*Feasibility of Storm Tracking for Automatic Control of Combined Sewer Systems*	November-78	Van-Thanh-Van Nguyen, M.B. McPherson & Jean Rousselle	The objective of this empirical study was to explore the feasiblity of tracking storms for combined sewer system automatic control applications. Data used were for the largest depth events in a 27 month record of hourly rainfall from a network of 18 gauges in the Montreal (Canada) region.
Journal of The Water Resources Planning and Management Division— *Urban Runoff Control Master Planning*	November-78	M.B. McPherson	The thesis of this paper is that rational planning requires conjunctive consideration of the quantity and quality aspects of urban runoff within a comprehensive, multiple-use framework. Explored are some of the more obvious arguments with reference to land-use factors, economic evaluations, flood control considerations, performance simulation, and metropolitan water resource inventories. There appear to be very few master plans extant that have integrated water quantity management with water quality management. Given the institutional constraints in metropolitan areas, perhaps the most that can be expected is adoption of an integrated, comprehensive, or systems approach to

REPORT TITLE	DATE PUBLISHED	AUTHOR(S) or EDITOR(S)	ABSTRACT
			urban runoff planning. However, such an approach is subtle in a documentary sense, and more rational conjunctive master planning may be in progress than external indications seem to imply.
Paper— *Reviews of Geophysics and Space Physics*	November-78	M.B. McPherson	Because the complex interactions of human activity in concentrated settlements with air, water and land must be taken into account, urban hydrology is a distinctive branch of the broad field of hydrology. As opposed to conventional hydrology, because urban development everywhere has been in continuous states of expansion and flux, urban hydrology contends with the dimension of dynamic change. Also, urban water management utilizes the social and biological sciences as well as the physical sciences. The fact that the term "urban hydrology" gained currency less than two decades ago reflects the lag in the recognition that America became an urban nation over half a century ago. Since then, the term has been tacitly expanded to include all urban water resource matters in, or interfacing with, the hydrologic cycle, including water quality considerations. Because the period 1975-1978 was dominated by attention to quantity and quality of urban runoff, that emphasis will be preserved here. However, one of the most important trends of the period was a growing acceptance of a perception of water in urban areas as a totality, as a resource [Kohlhaas, 1975; Dendrou et al., 1978a]. This integrated view is exemplified in a prognostication of needs for urban water models by Sonnen et al. [1976].
ASCE Urban Water Resources Research	November-78	Van-Thanh-Van Nguyen, M.B. McPherson &	Addendum 6 of the ASCE Program report "Urban Runoff Control Planning Addendum 5 of the ASCE Program report "Urban Runoff Control Planning," June

REPORT TITLE	DATE PUBLISHED	AUTHOR(S) or EDITOR(S)	ABSTRACT
Program TM-35—*Feasibility of Storm Tracking for Automatic Control of Combined Sewer Systems*		Jean Rousselle	1977
Urban Land Institute, American Society of Civil Engineers and National Association of Home Builders— *Residential Erosion and Sediment Control— Objectives, Principles and Design Considerations*	1978	Published jointly by Urban Land Institute, American Society of Civil Engineers and National Association of Home Builders	This report covers some of the basic concepts for protecting land and water resources against the detrimental impacts of erosion and sedimentation during residential construction. The objectives, principles, and design considerations are complex individually and collectively on a variety of levels; many detailed considerations have, of necessity, been omitted. The following are some of the basic concepts generally found to be useful: • Erosion and sediment movement and deposition have both beneficial and detrimental effects. • Sediment movements should not be permitted at rates or in quantities which will cause significant residual damage. Under ideal conditions any change in the nature or amount of sediment leaving a site as a result of construction should maintain or improve environmental quality when compared to pre-construction conditions. While these ideal conditions cannot always be acieved, the importance of environmental quality to man's long-term welfare and survival means that sound judgment must be exercised in establishing allowable rates of change from those expected in the natural cycle. • Strong emphasis needs to be placed on "natural" engineering and land

REPORT TITLE	DATE PUBLISHED	AUTHOR(S) or EDITOR(S)	ABSTRACT
			planning techniques, which will not only preserve and enhance natural features of the land, both on and off the site, but protect them. There are techniques which use and improve the natural processes taking place at a construction site, during and after the actual construction period, rather than ignoring or replacing them with artificial systems. • There must be increasing recognition that each site has its own set of natural resources, land use limitations, envirionmental conditions, and occupancy requirements. These factors and their inter-relationship vary from site-to-site within a community, and variations in design standards will be required for achievement of optimum off-site protection. • There must be continuing recognition of a balance of responsibilities and obligations between individual land owners and the public for the protection of the environment from the adverse impacts of excessive erosion and sedimentation. It must be understood that significant immediate and long-term expenditures for the construction and maintenance of this protection will be incurred by individual homeowners and the community. A balance must be struck in determining the acceptable ranges of damage, in order to avoid restricting housing availability and choice.
Paper—*Urban Water Balances*	Post 1978 (per references)	M.B. McPherson	Much of this presentation is necessarily a recapitulation of parts of three earlier papers: *Conservation in Household Water*

REPORT TITLE	DATE PUBLISHED	AUTHOR(S) or EDITOR(S)	ABSTRACT
Considering Conservation			*Use; Measures for Municipal Water Conservation,* and *Water Conservation: Response.* (Includes figures that are editorial modifications of illustrations prepared in 1975 and 1976 respectively, that have been reintroduced several times since.
A Study of ASCE Urban Water Resources Research Council (UWRRC) TM-36— *Introduction to Social Choice Theory for Environmental Decision Making*	February-79	Philip D. Straffin, Jr.	Chapter 2, "Power in Decision Making Bodies," and Chapter 3, "Voting Methods for More Than Two Alternatives," will be of immediate interest to anyone involved with environmental decisions. The author has crystalized the principles of relevant mechanisms that have heretofore been dispersed widely over the technical literature of the social sciences.
14th Annual Henry M. Shaw Lecture in Civil Engineering— *Challenges in Urban Runoff Control*	March-79	M.B. McPherson	Under Section 208 of the Federal Water Pollution Control Act of 1972, areawide planning for water pollution abatement management has been undertaken in a majority of metropolitan areas. Urban runoff considerations are an issue in most of these areas. Supported by a National Science Foundation grant, the ASCE Program assembled a report in 1977 intended to assist agencies and their agents preparing areawide plans, from the standpoint of major urban runoff technical issues in long-range planning. The report plus two new addenda were published in a single volume by EPA in 1978. Under a renewal grant from NSF, on-site elaboration of the principles presented in the 1977 report, and their adaptation to local conditions, have been provided in seminars, workshops and conferences hosted by areawide planning agencies. The

REPORT TITLE	DATE PUBLISHED	AUTHOR(S) or EDITOR(S)	ABSTRACT
			main points made at the first of such visits were summarized in a paper on *Urban Runoff Control, Quantity and Quality* presented at an APWA Urban Drainage Workshop, Omaha, Nebraska, March 14, 1978. The number of visits has since doubled, and the paper which follows is a substantial enlargement of the Omaha paper that takes into account the more recent visits and related experiences. The dozen visits made so far have been at widely separated urban centers across the country, ranging from Maine to Nevada to Texas to Florida. This ASCE-NSF project wil be concluded in a few months. A presentation for the Hydraulics group of the Boston Society of Civil Engineers Section of ASCE in Cambridge, Massachusetts, on January 31, 1979, afforded an opportunity to test the credibility of the more recent portions of this paper.
ASCE Urban Water Resources Research Program TM-37— *Challenges in Urban Runoff Control (Planning, Implementation & Simulation)*	August-79	M.B. McPherson	The ASCE program assembled a report in 1977 intended to assist agencies and their agents preparing areawide plans, from the standpoint of major urban runoff technical issues in long-range planning. Included with the report were two addenda on "Metropolitan Inventories" and "The Design Storm Concept." The report, plus two additional separately released addenda were published in a single volume in 1978.
Proceedings of Two Special Sessions, Spring Annual Meeting, American Geophysical Union,	September-79	Edited by M.B. McPherson	This Technical Memorandum is a supplementary report for a project supported by the NSF.

REPORT TITLE	DATE PUBLISHED	AUTHOR(S) or EDITOR(S)	ABSTRACT
Washington, DC, 28 May 1979 - ASCE TM-38— *International Symposium on Urban Hydrology*			
ASCE Urban Water Resources Research Program TM-39	October-79	M.B. McPherson	Proceedings of a Special Session, Spring Annual Meeting, American Geophysical Union, Washington, DC, May 29, 1979
Draft Paper— *Introduction to Digital Recording of Rainfall by Radar*	September-79	M.B. McPherson	The intended audience of this first draft is primarily the civil engineer interested in automatic control of either combined sewer systems or separate storm sewer systems or in flood warning systems for urban streams. It is widely believed that a telemetered rainguage network would be required for real-time control or warning. The most advanced applications would include predictions of rainfall on individual catchments, two or three hours in advance. The means for such rainfall prediction is digital radar recording, a capability that has emerged only very recently. Following is an attempt to summarize the state of the art in radar rainfall measurement and prediction from the viewpoint of potential applications in urban water resources operations. This draft has been prepared under support of USEPA Grant No. R806702010 awarded to ASCE for a project on *Feasibility of Automated Combined Sewer Systems Allied with Metropolitan Flood Warning Systems.* In its final form, it will be a chapter in the final project report, around mid-summer of

REPORT TITLE	DATE PUBLISHED	AUTHOR(S) or EDITOR(S)	ABSTRACT
			1980.
Draft Paper—d *Metropolitan Flood Warning Systems*	December-79	M.B. McPherson and William G. DeGroot	The role of the first author in the preparation of this first draft has been supported under USEPA Grant No. R806702010 awarded to ASCE for a project on *Feasibility of Automated Combined Sewer Systems Allied with Metropolitan Flood Warning Systems*. In its final form, it will be a chapter in the final project report, around mid-summer of 1980. Distributed earlier to EPA-ASCE project cooperators was *Introduction to Digital Recording of Rainfall by Radar*, draft of September 28, 1979 which will become another chapter in the final project report. The third major text chapter will be on *Automated Combined Sewer Systems*, and will feature findings from a study underway at Ecole Polytechnique of Montreal. It is generally agreed that complete automatic control of combined sewer systems will require the use of special new radar capabilities. In a given metropolitan area, such a special radar installation backed by calibrating telemetry-connected rainguages could also serve as part of a metropolitan flood warning system. A central thesis of the EPA-ASCE project is that the attractiveness of automatic control of urban runoff may be expected to increase as the territory served becomes larger and the number of functions served increases. Thus, the purpose of this presentation is to review prospects from a metropolitan flood warning system perspective, focusing on leading-edge initiatives in the Denver, Colorado area.
EPA Grant No. R806702— *Integrated Control of*	June-80	M.B. McPherson	Abatement of pollution from combined sewer overflows has been regarded by some localities as requiring the addition of extensive new storage, transport and

REPORT TITLE	DATE PUBLISHED	AUTHOR(S) or EDITOR(S)	ABSTRACT
Combined Sewer Regulators Using Weather Radar			treatment facilities to their systems. The central purpose of the study reported herein was to explore the technical feasibility of some less costly measures that might prove adequate for other communities with less stringent abatement requirements, namely the possibility of reducing the extent of overflows from conventional combined sewer systems that typically divert some stormwater via regulators to interceptor sewers. The means for increasing the abatement efficiency of such systems was postulated as the real-time, integrated operation of all interceptor regulators from a central computer. Dynamic regulators are normally actuated on the basis of local control whereby the interceptor flow stage in the immediate locality of a given regulator is maximized. When no in-line or other in-system storage can be called upon, no advantage over local automatic dynamic regulator control can be obtained by integrated operation of regulators unless expected flowrates to the interceptors can be estimated in advance of their actual occurrence, and unless an operational bias is introduced which either minimizes overflows from only some of the regulators on an interceptor or favors the timing of overflows from all of the outlets, such as during the initial storm period. Computations for an actual system covering a summer season indicate that, even when a time-varying pollutant parameter is accounted for, the reduction in mass of a pollutant entering the receiving water under integrated operation of regulators may be only marginal. Consideration is given to the potentials for exploiting in-line storage induced in collector and outfall sewers to gain greater flexibility with integrated regulator

REPORT TITLE	DATE PUBLISHED	AUTHOR(S) or EDITOR(S)	ABSTRACT
			operation.
			As part of an introduction to digital recording weather radar for non-meteorologists, newly available rainfall tracking, measuring and predicting capabilities of such radars are described. Use of these capabilities would be required to achieve maximum effective integrated regulator control.
			A thesis of the study was that the attractiveness of an integrated regulator control system would be limited if it was necessary to dedicate a radar mostly to that service. Accordingly, auxiliary uses for radars in metropolitan areas are given consideration, particularly for urban flood warning systems.
Conference Proceedings of the International Symposium on *Urban Hydrology*	May-Jun-83	A.R. Moodie, et al.	This Symposium, the second in what we expect will be a series of exchanges of information in urban hydrology on a national basis, was held as a follow-on to a symposium held in Washington, DC, in 1979, at which national reports on the current status of urban hydrology were presented. The participants agreed to meet again in 1983 to present national progress reports including assessments of what advances have taken place, what progress has been made, and what research needs remain unfulfilled. This Symposium is the result, and includes progress reports from all the original participants, as well as status reports from Japan, Venezuela, Switzerland, Finland, and Nigeria, who were not represented in 1979.
U.S. Environmental Protection Agency, Municipal Environmental	Arpil-81	M.B. McPherson	In this study, the possibility of reducing the extent of overflows from combined sewer systems was studied. In general, when no in-line or other in-system storage is used, integrated regulator operation has no advantage over local automatic-dynamic

REPORT TITLE	DATE PUBLISHED	AUTHOR(S) or EDITOR(S)	ABSTRACT
Research Laboratory, Research and Development —*Project Summary Integrated Control of Combined Sewer Regulators Using Weather Radar*			regulator control unless (1) expected flow-rates to the interceptors are estimated before they occur, and (2) an operational bias is introduced that either minimizes overflows from only some of the regulators on an interceptor or favors the timing of overflows from all of the outlets, such as after the initial storm period. A review of the capabilities of digital recording weather radar indicates it has the best potential for estimating rainfall needed for flowrate predictions. Other possible uses for such radars in metropolitan areas were considered, particularly their use as part of urban flood warning systems. The possibility of inducing in-line storage in collector sewers to gain greater flexibility with integrated regulator operation was also considered.
Proceedings of the Conference on— *Stormwater Detention Facilities*	August-82	L. Scott Tucker, Ben Urbonas, William DeGroot, et al.	The objective of this conference was to explore issues, define available technology, identify problems, and identify research needs with regard to the planning, design, operation, and maintenance of stormwater detention facilities. This was accomplished by bringing together the leaders in the field, and thereby merging the researcher, the technical practitioner, and the institutional practitioner to explore the issues. The conference concentrated more on the quantity aspects of stormwater detention than on quality. This is not to diminsh the importance of the quality aspects of detention, but, as a practical matter, it was necessary to limit the subject matter so that issues could be explored in depth. At the same time, quantity and quality issues cannot be totally separated. As a result, a meaningful portion of the conference was directed at the use of detention for stormwater quality enhancement.

REPORT TITLE	DATE PUBLISHED	AUTHOR(S) or EDITOR(S)	ABSTRACT
			The conference was organized into five major areas. These are: 1) Recent developments, 2) quantity issues, 3) quality issues, 4) institutional issues, and 5) research needs workshops.
Conference Proceedings— *Emerging Computer Techniques in Stormwater and Flood Management*	Oct-Nov-83	Stuart Smith, et al.	Conclusions of the conference were as follows: 1. The rate of price decrease of computer-compatible field instrumentation means much better temporal and spatial resolution is possible for model applications. 2. Models and design applications must adapt to this data environment (e.g. calibration will become a routine part of design). 3. Database management systems are critical for the analysis of the vast quantities of input data becoming available, and for the output from continuous-modeling systems. 4. Color graphics and interactive digital input devices together form the preferred man/machine interface. 5. Modular structured FORTRAN '77 (and updates) will remain our high-level programming language of choice. 6. Serious consideration should be given to UNIX-based operating systems (for 8-bit, 16-bit, and 32-bit and networked systems. 7. When acquiring a microcomputer, buy the cheapest system that will meet your immediate needs only, considering hardware, software, instrumentation, implementation, maintenance and design-time turnaround. 8. Buy immediately. 9. Undergraduate education must

REPORT TITLE	DATE PUBLISHED	AUTHOR(S) or EDITOR(S)	ABSTRACT
			adapt to the evolving personal microcomputer environment. 10. University researchers must be given state-of-the-art 32-bit personal computers with hard disks and color graphics (say about $50,000 each). This will ensure that research activities at the graduate level will be applicable when completed. 11. Engineers should receive better training in software engineering. 12. Utilities such as compilers, editors, and the like should be made more engineer-user oriented. The American Society of Civil Engineers should actively influence standards (this presumably would apply equally to the Canadian Society of Civil Engineers!) 13. Hardware and software is evolving too rapidly to hold conferences such as this only once every 3-4 years, especially over the next few years. The next conference should be held within one year at a central location. 14. Distribution of a conference tape is a good idea, and all speakers at the Conference should be required to contribute the software reported in their papers.
Proceedings of an Engineering Foundation Conference—*Impact and Quality Enhancement Technology*	June-86	Ben Urbonas and Larry Roesner, et al.	This book contains the papers presented at the Engineering Foundation Conference, "Urban Runoff Quality," held June 22-27, 1986 in Henniker, New Hampshire. Topics covered include data needs and collection technology, pollution sources and potential impacts on receiving waters, institutional issues, effectiveness of best management prctices, detention, retention, and wetlands. A workshop on research and future activities needs is summarized
Proceedings of	July-88	Larry	This book describes current practice in the

REPORT TITLE	DATE PUBLISHED	AUTHOR(S) or EDITOR(S)	ABSTRACT
an Engineering Foundation Conference on Current Practice and Design Criteria for Urban Quality Control— *Design of Urban Runoff Quality Controls*		Roesner, et al.	design of pollution controls for urban runoff. The contents comprise the proceedings of an Engineering Foundation Conference held in July 1988, titled *Current Practice and Design Criteria for Urban Runoff Water Quality Control*. The papers are concerned with the pragmatic, functional design and maintenance of devices that have been demonstrated to work in the field. These include: wet and dry detention ponds, infiltration devices, sedimentation tanks, swirl concentrators, wetlands, and source controls. Three related sessions address receiving water responses to urban runoff, the Water Quality Act of 1987, and institutional issues related to implementation of urban runoff quality control. These proceedings bring together much of the collective knowledge of American and European technology in the subject area. Many of the authors are internationally recognized in their field.
Proceedings of an Engineering Foundation Conference— *Urban Stormwater Quality Enhancement— Source Control, Retrofitting, and Combined Sewer Technology*	October-89	Stuart G. Walesh, et al.	These proceedings contain the papers presented at the Engineering Foundation Conference on *Urban Stormwater Quality Enhancement—Source Control, Retrofitting and Combined Sewer Technology*, held October 22-27, 1989, in Davos Platz, Switzerland. Session topics include institutional issues; stormwater regulations and standards; on-site detention, infiltration and percolation; inlet controls; planning; real-time control; in-systems control; treatment; rehabilitation; and research needs.
Proceedings of an Engineering Foundation Conference— *Stormwater NPDES*	August-94	Ben Urbonas, et al.	These proceedings *Stormwater NPDES Related Monitoring Needs*, consist of papers presented at the Engineering Foundation Conference held in Colorado, August 7-12, 1994. The Conference brought together 90 experts in the field of

REPORT TITLE	DATE PUBLISHED	AUTHOR(S) or EDITOR(S)	ABSTRACT
Related Monitoring Needs			urban stormwater management to discuss the current state of the U.S. Environmental Protection Agency's Non-point Pollution Discharge Elimination System (NPDES) regulations related to discharges of urban stormwater, and the monitoring requirements under those regulations. The objective was to summarize the current state of stormwater monitoring with respect to meeting these regulatory requirements, and to lay out an agenda for the future. Technical sessions included: 1) An overview of stormwater monitoring needs; 2) locating illicit connections; 3) system runoff characterization; 4) NPDES compliance monitoring; 5) policy and instutional issues of NPDES monitoring; 6) BMP monitoring for data transferability; 7) monitoring receiving water trends; and 8) stormwater and best management practice (BMP) monitoring. A major conclusion reached by the conferees was that existing monitoring requirements will not yield the information necessary to determine impacts on the environment or to evaluate the effectiveness of BMPs.
Proceedings of an Engineering Foundation Conference— *Effects of Watershed Development and Management on Aquatic Ecosystems*	August-96	Larry Roesner, et al.	These proceedings, *Effects of Watershed Development and Management on Aquatic Ecosystems*, comprise papers presented at the Engineering Foundation Conference held in Snowbird, Utah, August 4-9, 1996. The conference brought together 75 experts in the fields of environmental sciences, and engineering to discuss problems, solutions, and issues associated with the preservation, maintenance, and re-establishment of ecosystems in urban areas. Technical sessions included: 1) Approaches to expanding monitoring beyond water quality; 2) advances in using toxicity bioindicators in urban aquatic ecosystem assessment; 3) effects of watershed development n hydrology and aquatic

REPORT TITLE	DATE PUBLISHED	AUTHOR(S) or EDITOR(S)	ABSTRACT
			habitat structure; 4) impacts of watershed development of aquatic biota; 5) consequences of watershed development on stream morphology; 6) watershed development effects in arid and semi-arid regions; and 7) management and institutional issues. The papers and presentations demonstrate that management of urban water resources embodies a diverse set of scientific disciplines and engineering issues. But they also reveal that no template currently exists for integrating these disciplines and issues into municipal programs for urban water resource management. These proceedings do, however, establish the basis for building an integrated approach to development of sustainable urban water resource programs that embody the appropriate scientific and engineering considerations, along with the institutional structure to implement the program.
Proceedings—Sustaining Urban Water Resources in the 21st Century	September-97	A. Charles Rowney, et al.	These proceedings, *Sustaining Urban Water Resources in the 21st Century*, contain papers presented at the Engineering Foundation Conference held on Sepbember 7-12, 1997, in Malmo, Sweden. This conference brought together 90 international experts in fields including engineering, sciences, planning, land development, and landscape architecture to discuss problems, issues, and solutions associated with the development of stormwater management programs that truly sustain and enhance the urban water environment. Most technical sessions focused on case studies where a municipal or private development project resulted in clearly demonstrable improvement in an urban water resource. Other papers addressed the science and technology of urban runoff flow and quality management practices. These proceedings establish a

REPORT TITLE	DATE PUBLISHED	AUTHOR(S) or EDITOR(S)	ABSTRACT
			basis for development of sustainable urban water resources program that embody the appropriate scientific, engineering, cultural and institutional structures to implement the programs.
Proceedings of an Engineering Foundation Conference—*Linking Stormwater BMP Designs and Performance to Receiving Water Impact Mitigation*	August-01	Ben Urbonas, Jonathan E. Jones, et al.	*Linking Stormwater BMP Designs and Performance to Receiving Water Impact Mitigation*, consists of papers presented at the Engineering Foundation Conference held in Snowmass, Colorado, August 19-24, 2001. It brought together professionals of many disciplines to discuss and debate the linkages between various BMP designs and their performance and ability to mitigate receiving water impacts of urbanization. Specific areas addressed included urban watershed trends; regulatory and institutional perspectives; what is known about impacts of urbanization on receiving waters; BMPs and linkages to in-stream integrity; need for and examples of in-stream controls and habitat enhancements; policy issues related to zero and de-minimus impact development policies; design for sustainable water resources; information and monitoring needs to evaluate impacts mitigating potential of BMPs; and experience and science outside the United States.

C:\971-023\000frl\ASCE Combined Sewer Separation Project TM.doc

Subject Index

Page number refers to the first page of paper

Author Index

Page number refers to the first page of paper